FLOWCHARTS

FLOWCHARTS

by NED CHAPIN

AUERBACH®
publishers

princeton
philadelphia
new york
london

CONTENTS

LIST OF FIGURES

PREFACE

This book covers: program flowcharts, system flowcharts, the creation of flowcharts, computer-drawn flowcharts, the use of flowchart templates, ANSI Standard flowcharts, the history of flowcharts, and the interpretation and use of flowcharts. In addition this book includes a number of carefully drawn flowcharts that can serve as models.

Programers, analysts, managers of programers and analysts, users of computers who need to appraise how well the computer serves them, and students who use or are learning about computers—all will find much of value in this book.

This book does not try to teach programing, nor is it tied to any programing language. A little knowledge of any programing language will enable the reader to get more from this book. Any language will do about as well as the next— COBOL, FORTRAN, BASIC, Autocoder, Assembly Language, ALGOL, JOVIAL, MAD, or almost any other.

For the convenience of the reader who wants to delve into only a single subject (such as perhaps *system flowcharts* or *computer-drawn flowcharts*), the chapters in this book are designed to be relatively freestanding. To accomplish this, the author has represented definitions of some basic forms and terms, and some figures as well, in several ways in different places in the book. This will enable many readers to use this book more efficiently.

I invite you to make a note of your comments, reactions, and suggestions and to correspond with me about them. Please send your letter in care of the publisher. Your feedback will help make any future editions of this book more useful.

NED CHAPIN

FLOWCHARTS

FLOWCHARTS

SITUATION

Problem

People who work with computers need ways of describing the work computers do. The problem is not one of describing the work to the computer. For that, we use programing languages and job-control languages. The problem is rather one of describing the work for the benefit of other people. In short, the problem is people-to-people communication.

The person who knows what the computer does or is to do often must describe these things to other people. The topics of communication usually cover what the input data is, what the output data is, and how the computer transforms the input data into the output data. Often the topics also include both the operations done on data prior to the time it becomes input data and the operations done on the output data after it leaves the computer.

To facilitate such communication people give names to the operations computers can do and to the data the computer operates on and produces. This greatly helps to describe what the computer does. But operations done on data involve an element of sequence; that is, the same operations done in a different order may yield a different result, even when one starts with the same data. Consider the case of

adding and multiplying together three numbers, such as
3, 5, 7. Do we get the same result when we multiply 3×5
and then add 7, as when we add $3 + 5$ and then multiply
by 7? Are these the only possibilities? Hence in trying to
communicate about computer operations, people must also
usually include information about the sequence of the
operations.

A very practical question is: How can we describe these
operations in a way that is not only useful to the person
who wants to learn about the operations, but is also easy
for the person who is describing the operations? To a large
extent, this is a question of documentation; that is, it is a
part of the complete description of a computer's operation
and how people use this operation. Documentation, however, also concerns such things as how to run a computer,
what the formats of the data are, and why we want the
computer to process the data in this way. For these broader
and more general questions, documentation is needed to
provide adequate answers.

Full documentation, however, is not needed for the more
restricted questions such as: Which data serve as input?
Which data does the computer produce as output? What
operations does the computer apply to produce the output
data from the input data? A shorter answer can serve—but
what kind of answer?

Criteria

In seeking an answer, people have tried a number of
different techniques. From experience we have come to feel
that some techniques are more useful and serve better in most
situations than do others. From that experience people have
developed informal criteria to help them judge the convenience of describing the operations computers do.

The most important of these criteria are:

1. *Is the technique easy to use?* Effective communication

requires a facile means for exchanging information between one person and another. Something that is laborious, involved, or tortuous breaks down in practice and fails to serve well.

2. *Is the technique quick to use?* Something that is time-consuming typically breaks down in practice. In working with computers, people frequently are under the pressure of deadlines. They do not have the time for a slow technique. The best techniques, therefore, are those which allow the computer to produce outputs by itself with a minimum of assistance from people.

3. *Is the technique simple?* To be of value, it should have simple rules and few component parts. It should not require a lot of study, practice, or experience to use.

4. *Is the technique scanned?* A person who wants to get a quick, general idea does not want to be delayed by too much information. Whatever the technique is, it should enable its user to identify highlights, major features, or the general pattern of what is being described. For this, most people find that graphic techniques permit the easiest scan.

5. *Is the level of detail controllable?* This is important, both for the person who produces the technique and for the person who uses it. Each must be able to select and work at the level of detail that is important for him. If the level the user wants is not the one selected by the producer, there must be some way for the user to change the level. Here we come to a practical problem. It is usually impossible for the user to get more detail out than the producer built in, but it should always be possible for the user to be able to work at a more general level of detail than the producer provides.

6. *Is the technique free of ambiguity?* This is a question of degree and may depend on the level of detail. Whatever the technique is, it can be related on a one-to-one basis with significant features of what is to be described. The technique should permit no ambiguity in the essential questions of which data, operations, and sequences are necessary or desirable.

7. *Is the technique independent of the computer and of the computer language?* Communication is improved when people do not require a knowledge of a particular computer or of a particular programing or job-control language in order to use it. Moreover, this independence facilitates changing work and staff from one computer to another as work loads and computer availabilities change.

SOME SOLUTIONS

The need to describe computer operations is one of long standing. It was recognized early in work with computers; people cast about for solutions and tried a number of different techniques. A quick review of some of the major techniques may help to account for the popularity of one of them— flowcharts.

Formulas from mathematics is one technique. Their use is and has been favored by many. In some situations, most frequently in computational work, formulas are the technique of choice.[1] In general, however, mathematical formulas suffer from several difficulties: (1) the computer does more operations than the formula indicates directly; (2) usually the sequence of operations is not free of ambiguity; (3) the level of detail is fixed; and (4) the formulas are not easily scanned with good comprehension.

Written descriptions using words is another technique. This has been used extensively to describe computer operations. Because the descriptions are conveniently produced and easily read, many people have learned how to scan English language texts. Written descriptions can also be applied to a much wider variety of situations in which

[1] See, for example, Daniel D. McCracken and William S. Dorn, *Numerical Methods and FORTRAN Programming*, New York: John Wiley & Sons, Inc., 1964, Chap. 3. See also note 6.

people use computers, than can mathematical formulas. But such text descriptions are verbose; the order of description need not have any relationship at all to the actual sequence of operations, and ambiguity is common. These are major shortcomings.

Programing languages is the third technique for describing computer operations. This seems a natural use for programing languages since we use them anyway for telling the computer what it is to do. The imperatives and declaratives of a programing language can be read as the description of the operation.[2] Although this supplies a fine one-to-one correspondence with the operation being performed and shows sequence clearly, there are some difficulties. Programing languages cover well what goes on inside the computer, but they are difficult to use in describing both the preliminary operations on data prior to the time they become input and the subsequent operations on output data. Programing languages also tend to be too detailed and too verbose for human convenience. They are fixed in their level of detail and often are relatively difficult to learn to use.

Process charts are one of several graphic techniques from industrial engineering used for describing computer operations.[3] They are very effective in identifying who does an operation and can provide a way for describing what the operation is. They are also particularly useful, both for operations prior to the time data become input and for subsequent operations on output data. But they tend to be weak in describing what operations are performed in the computer itself in spite of their explicit recognition of the importance of the sequence of operations. Further, the use of process charts is not widely known, even though it is fairly easily learned.

[2] A classic example is H. Perstein, "Condensed Syntactic Description of JOVIAL (J3)," *SICPLAN Notices,* vol. 1, no. 7 (July 1966), pp. 13-15.
[3] Several varieties are shown in Robert N. Lehrer, *Work Simplification,* Englewood Cliffs, N.J.: Prentice-Hall, Inc., 1957.

Some graphic techniques, such as document flowcharts, are useful for describing computer operations.[4] These have much to recommend them, particularly for describing both operations on data prior to the time they become input and subsequent operations on output data. Their emphasis on the sequence of operations is strong, but they tend to be weak on operations within the computer itself. They are easily scanned and can be used with several levels of detail.

The five techniques just listed are all old. They were all tried early and are still in use. Since computers appeared, other techniques have been developed in an attempt to find a better solution to the problem of describing computer operations. Among these are three techniques worthy of special note: flowcharts, decision tables, and abstract languages.

Abstract languages are special symbols developed specifically for the purpose of describing operation on data in precise terms.[5] As such, they seem to be ideally adapted to meet the problems people have in communicating about computer operations. So far, however, they have enjoyed but little use. The major objection to them, apparently, has been their strangeness, which contributes to their difficulty in being learned.

Decision tables by contrast are used much more extensively than abstract languages. In practice, they are used more often than process charts and the document flowcharts mentioned earlier.[6] Decision tables clearly enumerate the operations performed. They can also identify the sequence of the operations, irrespective of whether the operations

---

[4] Several forms are shown in T. Radamaker, ed., *Business Systems,* Cleveland, Ohio: Association for System Management, 1963, Chap. 5.

[5] Some citations of the literature can be found in Ned Chapin, "A Deeper Look at Data," *Proceedings of the 1968 ACM National Conference,* Princeton, N.J.: Brandon/Systems Press, Inc., 1968, pp. 631-638; and R. L. Tucke, T. W. Miller, and R. G. Koppang, *Iverson Form and the Iverson Translator (Douglas Paper 5028),* Los Angeles, Calif.: McDonnell Douglas Corp., April 1968, 38 pp.

[6] Ned Chapin, "An Introduction to Decision Tables," *DPMA Quarterly,* vol. 3, no. 3 (April 1967), pp. 2-23.

are to be done within the computer or externally. While decision tables are easily learned and easily scanned, they are not usually easily produced, a characteristic they share with mathematical formulas. However, for logically complex situations, where mathematical formulas typically break down, decision tables are very capable in describing operations.

The flowchart is a graphic technique specifically developed from previously existing graphic techniques for the purpose of representing computer operations. It is fairly easily produced and fairly easily learned, having only a few relatively simple rules and few component parts. It can be used to describe unambiguously the way computers handle data. It can also be used to represent operations on data prior to the time they become input and after the time they are produced as output, as well as represent operations done within the computer. Because the flowchart is a graphic technique, it can be scanned quickly. Further, it can be read at almost any level of detail.

ROLE OF FLOWCHARTS

Because flowcharts meet the criteria well, people commonly use them for describing work done or to be done by computers in a number of different circumstances. In fact, people have found flowcharts so convenient that they utilize them in most of the phases normally observed in the preparation of work for a computer: problem definition, system analysis, system design, programing, debugging, conversion, documentation, operations and maintenance. A brief look at the role of work descriptions in flowchart form can indicate some of the reasons people find flowcharts so convenient.

The problem-definition phase concentrates on what the computer is to do. What function is it to perform in the system? How is the output going to be useful? What input will be required to produce the output? Unlike the later phases which are concerned with these same questions in much detail, the problem definition phase deals with them in general terms. For that purpose, the description technique needed is one for recording general descriptions and data identifications. Flowcharts can serve to state some of the primary conclusions developed from the problem-definition phase.

The system-analysis phase attempts to break down into component parts the important things affecting the way a computer may do a job. It looks at the requirements the operation will have to meet, the timing considerations, the operational limitations, the data that are available, and the cost of alternatives. In this phase, the flowchart produced in the preceding phase can serve as a general organizing tool. As a result of the phase, a more elaborate and more detailed description of the job may emerge. Usually this is not a flowchart.

The system-design phase, by contrast, attempts to produce, from the results of the system analysis, a comprehensive statement of the way a computer might be used to perform a job. This phase is synthetic in character, attempting to set forth procedures, sequences, and data identifications in detail. As such, one of its major results can be expressed in terms of a description of operations on data, their timing, and their sequence. For this, a flowchart can usually serve well.

The programing phase then must develop the details needed to provide directions to the computer. As such, programing uses the flowchart resulting from the system-design phase as a major source of information, but it may also produce more detailed flowcharts reflecting the actual implementation accomplished.

The debugging phase is the process of finding and correcting the mistakes made in the prior phases. Thus it often refers back to the flowcharts already developed. It may have to correct them or elaborate them in order to clarify ambiguities and fill in missing details. As such, debugging both uses flowcharts and produces flowcharts.

The conversion phase is the process of completing the implementation of a job. It consists of putting the job into routine production. It is concerned with developing and refining the procedures for handling the data that will serve as input, as well as the procedures which are to be performed on the output. As such, conversion uses as its starting point the flowchart descriptions of the operations that result from the previous phases. It attempts to make the actual operations that are to be performed match the planned operations as closely as possible. Its result is often a revised flowchart showing the details of the implementations.

The documentation phase ideally does not come all at once. It is best carried on throughout all the phases just noted. If documentation is defined, in part, as the production of a description of the work the computer is to do, then documentation is a key element resulting from, and contributed to, by each of the phases. In practice, however, documentation is often postponed to about the time of conversion. This enables a systematic review and summary to be made that incorporates the last-minute revisions.

The operation phase is the routine production of output. It typically consists of carrying out the work description embodied in the flowchart that was established during the preceding phases. As such, this phase rarely contributes to or augments that description although it often makes reference to it as a guide in the day-to-day execution of the work.

The maintenance phase, by contrast, consists of revising the work that is done. This, in turn, requires a revision of the description of the work and is often done, in part, by revising the flowcharts. For these reasons, this phase often recapitulates with changes many of the prior phases.

Maintenance, therefore, typically uses flowcharts and also produces revised flowcharts to reflect the new way the work is done.

In all of these phases, the contribution made by flowcharts is as a means of communicating ideas from one person to another. This is because the flowchart provides a written record identifying the data and the sequences of operations.

ABOUT THIS BOOK

The objective in this book is to present flowcharts and to explain their use. It covers theory, standards, conventions, good practices, interpretation, and computer-produced flowcharts. It explains how to create and draw flowcharts, and it includes some exercises to sharpen the reader's skills.

The author assumes that the reader knows something about automatic computers and programing, but does not assume that he is an expert regarding either. This book does not attempt to teach programing, nor is it slanted in favor of either computational (mathematical) or administrative uses of computers. Most of its examples are taken from simple utility and statistical problems, since these occur both in computational and administrative uses of computers. Such examples might include making a copy of data and figuring the average. Further, this book is independent of, but usable with, any programing language, such as COBOL, FORTRAN, BASIC, Assembly Language, ALGOL, Autocoder, Easycoder, APL, LISP, JOVIAL, MAD, APT, PL/1, and others.

Lastly, this book does not limit itself to flowcharts of program algorithms for execution by a computer. It also treats flowcharts for systems, including operations on data not performed by a computer. For both program and system flowcharts, this book stresses the things the user needs to get flowcharts to serve his needs.

FLOWCHART FORMS

HISTORY

Early Developments

The intellectual father of flowcharting is John von Neumann. He and his associates at the Institute for Advanced Study in Princeton, New Jersey were the first to use graphic aids systematically for this purpose and to publish information about their use.[1] Even though the details of flowcharting today differ considerably from what these men advocated, the spirit, philosophy, and rationale of flowcharting remain much as they presented them.

For its own internal purposes and for dealing with customers, each of the major computer manufacturers has, over the course of the years, developed, adopted, published, modified, and advocated flowcharting conventions.[2] These have differed from vendor to vendor, in part as a deliberate attempt to distinguish one vendor from the other competing vendors, and in part as a sincere attempt to reflect what

---

[1] H. H. Goldstine and John von Neumann, *Planning and Coding Problems for an Electronic Computing Instrument*, Princeton, N.J.: D. Van Nostrand Co., Inc., 1947, vols. I, II, and III. The earliest unpublished use is reported by Fred Gruenberger to be a November 1946 letter from John von Neumann to Ed Paxon of the RAND Corporation.

[2] IBM Corporation, *Flowcharting Techniques*, C20-8152, New York: IBM Corp., 1964; idem, *Problem Planning Aids, IBM Type 650*, New York: IBM Corp., 1956.

each has felt to be unique differences in its philosophy and approach to information-processing problems.

Users of computers have individually and collectively made decisions on flowcharting conventions. Most small and medium users of computers, and many large ones as well, have adopted the conventions presented to them by the vendor of the computer they have selected. A few larger users, however, have chosen to go their own way and develop their own internal standards. The United States Air Force, for example, has developed its own standards for the purpose and in practice has had a multiplicity of standards.

Users of computers acting collectively through the user groups have sometimes addressed themselves to the problem of standards for flowcharting. Since these user groups normally have been composed of users of only one vendor's computer, the effect usually has been to recommend modifications or suggestions to the computer vendor for changes in the vendor's standards. However, a few users groups, such as SHARE, have independently presented their own standards and have advocated them for general adoption.[3]

Individuals with competence and standing in the computer field have for their own use sometimes deviated from the practice of the various computer vendors and from the other sources. They have presented their own recommendations for flowcharting. These recommendations have been effective as a result of their authors' publications.[4] Some of these recommendations, however, have ceased to be significant with the arrival of formal standardization.

The use of the computer itself to produce flowcharts

---

[3] W. Barkley Fritz *et al*, "Proposed Standard Flowchart Symbols," *Communications of the ACM*, vol. 2, no. 10 (October 1959), pp. 17-18; and Mandalay Grems, "Comparison of flow chart symbols," *Communications of the ACM*, vol. 3, no. 3 (March 1960), pp. 174-175.

[4] Ned Chapin, *An Introduction to Automatic Computers*, Princeton, N.J.: D. Van Nostrand Co., Inc., 1963; Kenneth E. Iverson, *A Programming Language*, New York: John Wiley & Sons, Inc., 1962; Daniel D. McCracken *et al*, *Programming Business Computers*, New York: John Wiley & Sons, Inc., 1959.

has had relatively little influence on the development of flowcharts although programs for doing this have been in use since 1957, as a later chapter in this book indicates. Of more practical significance in popularizing the vendors' positions has been the vendors' practice of providing plastic templates as an aid in drawing flowcharts.[5]

Standards for Flowcharts

During the 1960's a committee attempted to develop a standard for flowcharting. It worked through the Business Equipment Manufacturers Association (BEMA) and the American Standards Association (ASA) with committee members drawn from computer vendors and a few major computer users. After the typical compromises, this committee drew up a proposed standard and circulated it for reaction. With revisions, it was approved in 1963 and published as American Standard X3.5. The Association for Computing Machinery (ACM) and other groups published this X3.5 Standard in their periodicals, giving it considerable publicity.[6] This effort toward standardization in the United States paralleled a similar effort conducted for the International Standards Organization (ISO).

Subsequently, in 1965 and again in 1966, 1968, and 1970, the X3.5 American Standard was revised. The 1965 revision was a major one and was published by the Data Processing Management Association (DPMA), but the 1966 and 1968 revisions were only minor ones.[7] The 1970 revision extended

---

[5] A recent example is IBM Corporation, *Flowcharting Template*, *X20-8020*, New York: IBM Corp., 1969, one plastic cutout drawing guide in a printed envelope.

[6] See, for example, "Proposed American Standard Flowchart Symbols for Information Processing," *Communications of the ACM*, vol. 6, no. 10 (October 1963), pp. 601-604.

[7] ANSI, "ASA Standards for Flowchart Symbols," *Journal of Data Management*, vol. 3, no. 11 (November 1965), pp. 18-22. This gives the 1965 version of the Standard

the X3.5 Standard to match more closely the ISO Standard.[8] In 1965 the American Standards Association changed its name to the United States of America Standards Institute, and the ASA X3.5 Standard was known as the USASI X3.5 Standard. Then in 1969, the Institute changed its name again, this time to American National Standards Institute (ANSI, usually pronounced "ann-see"). The flowchart Standard became the ANSI X3.5 Standard.

The flowchart Standard is not the only graphic aid that has gone through a formal standardization process and that can be used to describe operations on data. A Standard (X2.3.4) has also been adopted that is important for describing clerical procedures, such as those used in systems and procedures work.[9]

The graphic aids of that Standard are quite different from those discussed in this book on flowcharts. Logic designers of hardware also use graphic aids for stating the character of the machines they design for handling data. These graphic aids too have been the subject of a Standard (Y32.14) [10] and are quite different from flowcharts.

All the flowcharts presented in this book have been prepared to be in conformity with the ANSI X3.5 flowchart Standard. Moreover, the descriptions, directions, and suggestions in this book for drawing flowcharts also conform to the spirit and the letter of that Standard. Wherever the material presented by this book is not in conformity with the Standard, the nonconformity is identified as such, and some explanatory comment is provided.

Nonconformities may be of three types: a deviation, an augmentation, or a violation of the Standard. A deviation from the Standard refers to something that has an ambiguous

[8] ANSI, *Standard Flowchart Symbols and Their Use in Information Processing (X3.5)*, New York: American National Standards Institute, 1971.
[9] ANSI, *Standard Method of Charting Paperwork Procedures (X2.3.4)*, New York: American National Standards Institute, 1959.
[10] ANSI, *Standard Graphic Symbols for Logic Diagrams (Y32.14)*, New York: American National Standards Institute, 1962.

or equivocal relation to the requirements of the Standard. Different interpretations of the requirements of the Standard and of specific instances of the use of the Standard commonly give rise to deviations. Some examples of deviations are cited later in this book.

By contrast, an augmentation of the Standard is an elaboration or extension of the Standard to cover a situation not provided for. Since the Standard, of necessity, is silent on these points, those who use the Standard, in an attempt to make it fit their own needs better, try to stretch or add to it. For example, a particular kind of operation not specifically cited in the Standard might be represented by a particular flowchart form unlike anything already in the Standard. An example of augmentation is presented later in this book.

A violation of the Standard, by contrast, occurs when the Standard is directly contradicted by the user; that is, the Standard says one thing, and the user chooses to ignore the Standard and do something instead that is directly in opposition to it. An example of a violation of the Standard is described later.

It should be noted that a violation of the Standard is undesirable. An augmentation, by contrast, may be a very justifiable, or even a desirable, change in the Standard. A deviation falls in between and is usually neutral in its effect. Generally, deviations should be shunned, for their presence and use impair the communication value that standardization attempts to foster.

DEFINITIONS

Flowchart

A flowchart is a means of portraying, in graphic form, a sequence of specified operations performed on identified data. The identification of the data is usually by name, and the specification of the operations is usually by name or by type. Also included in some flowcharts is an indication of the media or the equipment which handle the data, or the means used to accomplish the operations on the data, or both. Figure 2-1 provides two examples of flowcharts.

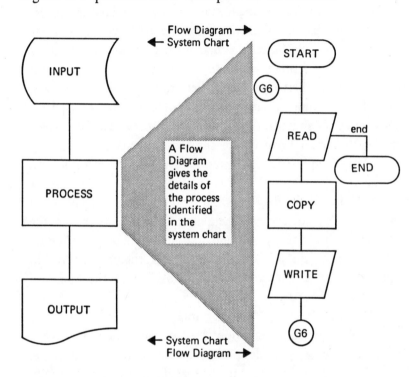

Figure 2.1. Two types of flowchart: the system chart and the flow diagram

- Do-It-Now Exercise 2.1. Write a verbal description of the data handling presented in the system shown in Figure 2-1. Then do the same, separately, for the program. Notice that you can understand a lot of what these flowcharts present, even though this book has not yet told you about them specifically.

The graphic part of a flowchart is composed of *symbols*, *outlines*, or *boxes* of various shapes with connecting *symbols*, *lines*, or *arrows*. The ANSI Standard prefers the term *symbols* to refer to this graphic part. This terminology suffices when people speak only of the graphic part of a flowchart, but it becomes ambiguous when people also need to speak of the names of data and of operations. People use symbols for these, too, but not symbols as meant by the Standard.[11]

To avoid such confusion, this book refers to the graphic part of a flowchart as being composed of *outlines*, where an arrow or line is also considered to be an outline. Following common practice, this book uses "symbols" to refer to the means of naming data and operations. Thus in Figure 2-1, the operation of copying is specified by the symbols COPY, which appear within a rectangular outline.

Types of Flowcharts

In common practice, flowcharts are of two types, types which reflect the two most common situations in which people use flowcharts. One situation is the representation of algorithms, especially those for execution by a computer. The other is the representation of systems without indicating the character of the component algorithms. The flowchart used for one situation does not serve well for the other. To avoid confusion in this book, therefore, a *flow diagram* will designate a flowchart of an algorithm, and a *system chart* will designate a flowchart of a system.

This distinction between the flow diagram and the system chart is vital because they serve in different roles. In the system chart, the focus is upon the inputs and the outputs produced by the sequences of runs, programs, or procedures. In contrast, the focus in the flow diagram is on the sequences of data transformations needed to produce an output-data structure from an input-data structure.

The flow diagram stresses *how;* the system chart stresses

[11] ANSI, "Working Paper: Graphic Symbols for Problem Definition and Analysis," *Communications of the ACM*, vol. 8, no. 6 (June 1965) pp. 363-365.

what. Whereas a system chart identifies programs, runs, or procedures by name and data structures by name, the flow diagram identifies individual operations on portions of data structures. The flow diagram is usually an elaboration of what is indicated by a single process outline in a system chart (see Figure 2-1). Typically, the system chart uses a greater variety of graphic outlines, but its logical complexity is relatively low. Typically, the flow diagram is more logically complex, even though the number of different graphic outlines utilized is fewer.

Other terms are also current in the field for these two situations. Thus other terms sometimes used for flow diagram are *block diagram*, *logic chart*, and *process chart*, as well as *flowchart*. For system chart, the other terms used are *run diagram*, *procedure chart*, and *flowchart*.

The stress in this book on system charts and flow diagrams does not mean that other types of flowcharts are nonexistent or illegal. On the contrary, people, through the course of the years, have developed many variations in the use of flowcharts. Usually these have been developed to meet particular situations, concentrating on certain aspects of the way computers and people interact with data. The fact, therefore, that variations exist which are not covered in this book does not mean that such variations of flowcharts are wrong or that they are not useful. It only means that they are not as widely used as the system chart and the flow diagram.

TEMPLATES

Purpose of Templates

From the point of view of mechanics, people usually proceed to draw a flowchart in one or two ways. One way is simply to grab a pencil or pen and scratch out the graphic outlines in freehand form. If one is working at a chalkboard or on scratch paper, this way is eminently satisfactory.

When neater, more easily read flowcharts are needed, people find it hard to draw freehand outlines with sufficient precision and neatness without spending an inordinate measure of time in the drawing process. To speed the drawing, while retaining the needed precision and neatness, people use templates. These serve as guides in the drawing of the outlines for a flowchart. Templates generally are made of plastic, with cutouts around which a pen or pencil can trace to produce the desired shapes.

Templates for this purpose are available from a wide variety of sources. Traditionally, the computer vendors have made them available free of charge to their customers and prospective customers. More recently, some of the vendors have been charging for these templates although some are still given away for goodwill.

Templates are also available from some manufacturers of plastic drawing aids, such as K&E, C–Thru, Plastigraph, and a number of others. These templates usually are unlike those available from the vendors in terms of their selection of outlines and general format. Some major users, dissatisfied with the templates available, have prepared their own. For example, the Royal Canadian Mounted Police have a flowcharting template of their own design for use in their data-processing operations.

Features of Templates

A good flowchart template should be almost transparent, but it should not be colorless. A colorless template is difficult to see and, therefore, is easily lost and misplaced. For this reason, templates are usually tinted, the common colors being gray, light green, light blue, and light yellow.

In addition, a good template has well-placed registration lines printed on it to assist in lining up the outlines neatly as they are drawn. The openings through a good template are free of rough edges and burrs, and all the edges are smooth. The straight lines are straight, and the curved lines turn smoothly on a good template.

A good template has a good registration (agreement) between the printing on it and the openings used for drawing the outlines. The cost of making a perfect registration is so high that perfection is achieved only by accident. As a practical matter, a tolerance of about half a millimeter (around $\frac{1}{64}$ inch) should be regarded as still an acceptable registration. A registration off by more than a millimeter (around $\frac{1}{32}$ inch) can cause some difficulty in using a template.

A good template is durable and typically is slightly flexible. It should not shatter or dent when hit or dropped, and it should not craze under repeated gentle flexing. The slight flexibility assists in moving the template. Ideally templates should not be thicker than about one-tenth of an inch and not thinner than about one-fortieth of an inch. The thicker templates are sturdier, are easier to handle, and can be used to draw neater flowcharts.

A good template has beveled openings or stands away from the surface on which it rests. The smallest part of an opening should be on the top of the template; that is, the bevel should come from the reverse, or back, side of the template. This has several advantages. First, it makes the template much easier to pick up by providing space for the fingers to grip it more easily in the openings. Second, it reduces smearings when ink or a ballpoint pen are used with the template. If the template is not beveled, the user can produce a similar effect by sticking on special lugs with pressure-sensitive adhesive, or by sticking small narrow strips or spots of masking tape (usually a double thickness) on the bottom side of the template. The locations for the lugs should be chosen so that they do not interfere (horizontally and vertically) with the registration marks printed on the template. The purpose of these lugs is to raise the template off the paper. Without this or a bevel, the ink may run between the template and the paper by capillary attraction and cause a smear.

A good template also offers several sizes of openings for each of the outlines included. The openings should be accurately sized so that the outlines drawn with the template conform not only to the desired ratio of width and height, but also to the general configuration. The ANSI issues a certification for some of the templates that meet this requirement. A good template may have rules or scales marked along some or all of its edges. Generally, these are marked in tenths, sixths, eighths, or sixteenths of an inch, or in millimeters.

Finally, a good template should have a smooth finish on all sides. This helps avoid picking up dirt from the fingers and paper and thereby decrease transparency of the template. The printing on the template should be recessed or engraved into the template on the bottom or reverse side, or laminated into the lower center portion of the template, to protect the printing from wear. The lettering and rules should not be susceptible to rubbing thin or being scraped off inadvertently in the course of normal use.

Ideally, the plastic material from which the template is made should be relatively free of static electricity. Unfortunately, with most plastics, static electricity cannot be avoided completely. If static electricity buildup is a problem, then one of the antistatic electricity compounds used by hi-fi enthusiasts for cleaning phonograph records can be used for cleaning the template. Cleaning the template as you would a pair of eyeglasses is also sometimes effective.

The Use of Templates

In using a template to draw flowcharts, one usually takes the steps listed below. It should be stressed at the outset that a balance must be struck between the degree of accuracy in the drawing of the outlines and the speed with which the drawing is done. In some work, such as work by a draftsman for reproduction, accuracy may come before speed. In cases where great accuracy is desired, templates cannot be used. For such situations, the outlines must be drawn from specifications. Chapter 7 of this book summarizes the specifications. Usually, however, minor inaccuracies in placement, alignment, and neatness can be tolerated in the interests of a speedy preparation of the flowchart with a template.

The steps are as follows:

1. *Select the shape (configuration)*. This is a problem of locating on the template the opening that can be used to produce the configuration you desire to draw. Since each different template typically has the openings in different places, no general guide can be offered. It is mostly a matter of becoming familiar with the template you have.

2. *Select the size*. Many templates permit drawing only one size of an outline. A few templates offer a selection of sizes. The usual procedure is to choose the smallest size that will just encompass the identifying symbols that are to be entered within the outline. Since some outlines must be drawn in two parts (with a movement of the template in between), attention may be needed to get the two parts to be of the same size.

3. *Align by the visual center*. To keep a flowchart looking neat, one should arrange the outlines so that their visual centers are positioned in a regular pattern—usually either vertically or horizontally. All of the flowcharts shown in this book were prepared observing this step. Here is where the horizontal and vertical alignment printing on the template proves its worth. Where no alignment marks are provided, or where only a general ruled background or set of grid lines is provided, the user must locate for himself an indicator of the center for each outline he wishes to draw.

4. *Select the size of the separating space*. This is the distance between the outline you now wish to draw and the adjacent outlines. Most people try to keep this space fairly

uniform; that is, each outline is equally distant from every other outline vertically and horizontally, but especially along the line of flow (the line of connecting arrows). Yet this need not be the case. Often it is desirable to provide additional space in order, for example, to improve the readability. Sometimes, too, it is desirable to crowd a flowchart on a page to make the end of the page coincide with a natural break in the flow, such as the end of a routine.

5. *Position the template.* Be sure the top of the opening is in the correct place as you put the template on the paper. In positioning the template, you should not slide it along the paper. Rather, it should be lifted clear and then set down on the paper as close to the desired position as possible. Sliding the template along the paper can smear ink and pencil lines that may already be on the paper. Once the template is positioned in general, a slight repositioning may be necessary to make the final alignment and spacing.

6. *Trace around the opening.* The best practice for drawing the outline, once the template is in position, is to hold the pencil or pen directly upright. Ideally it should be at right angles to the paper all around. Then, starting at a corner of the outline or at a point where the outline has a change in direction, move the pen or pencil at a uniform rate around the edge of the opening in the template. To keep the line straight, you should not tip the pen or pencil, and you should use a light uniform pressure to maintain the pen or pencil against the edge of the opening in the template. If the template is to be kept from shifting during this process, it usually must be held down with a light pressure.

When a pencil is used with the template, some of the carbon "lead" will usually rub off from the pencil onto the edge of the template. Sometimes it will fall like dust onto the paper. It is best to blow or lightly brush the dust away to avoid having it cause a smear later. It is often helpful to wipe with a tissue or cloth the interior of the opening used in order to remove the accumulated lead from the pencil.

When a ballpoint pen is used, the ink tends to build up just above the tip of the ball on the side of the pen away from the direction of motion. For example, if you are moving the pen from left to right, the ink accumulation will occur on the left-hand side of the pen. Then at a corner, this ink can drop down onto the paper and result in a blob or smear within half an inch (a centimeter) from the corner. Most commonly, this happens when you attempt to retrace a line in the opposite direction. When you go around an arc or a circle, the ink accumulation is distributed and tends to be self-cleaning after roughly half a circle. This is the reason why the ink buildup normally is not bothersome when you write with a ballpoint pen.

If fluid ink and drawing pens are used, even more care than with ballpoint pens must be exercised to keep the ink away from where the template touches the paper. If the ink touches these points, it may flow by capillary attraction into the space between the template and the paper, causing a smear. This is particularly a problem when it is necessary to reposition the template in order to draw the next ouline. The smear can occur when the template comes in contact with the damp ink.

7. *Remove the template by lifting it straight up.* Keeping the template clear of the work area makes possible a better perspective on the flowchart as it is drawn and helps to avoid smears.

8. *Work quickly.* After reasonable precautions have been taken to achieve the neatness called for in the work at hand, the objective is to work as quickly as possible to accomplish the drawing. Drawing flowcharts is much like writing computer programs: better something usable done today than something more elegant not done until next week.

Template Suggestions

Suggestions on drawing flowcharts with the aid of templates depend in part on the specific templates available. The suggestions that follow are generally applicable to most of the available templates. But to help the reader master the use of whatever templates he has available, the suggestions here reflect the use of templates of the most advanced design.

Two templates are described in this section. One is marked "program flowchart template." The other is marked "system flowchart template." Together the two templates permit one to draw all of the outlines needed for flowcharts. For program flowcharts, only the program-flowchart template is needed. For system charts, both templates may be needed. Some features of these templates are worth special note.

Along the edge of the templates are scales in tenths, eighths, and sixths of an inch, plus scales in millimeters and centimeters. The openings in the templates are beveled, with the flair-out of the bevel at the bottom.

The templates have provision for drawing outlines of different sizes by tracing around the inside or outside edges of the openings. The program-flowchart template provides for four different sizes for each outline, except for circles, for which the template has seven different sizes. The system-flowchart template has two sizes for each outline, except for punched tape, where only one size is provided.

To help maintain a neat appearance in the outlines, the person drawing them should observe a few simple rules:

1. Use as a writing tool a good pen or moderately sharp pencil (felt pens do not work well with templates).
2. Hold the writing tool straight upright.
3. Press the tip of the writing tool only very lightly against the template edge.

4. Draw the line at a fairly uniform speed, stopping smoothly as the tool approaches a corner.
5. Hold the template lightly in place to prevent it from shifting position as the line is drawn.

Rule 4 is especially important to observe when using the inner part of the template openings, as is shown in the right-hand portion of Figure 2-2.

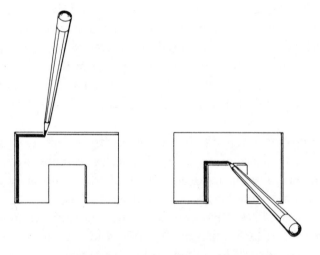

Figure 2.2. Drawing two sizes of a rectangular outline

If you are to get this number of outlines and sizes from a conveniently small template, the template must be repositioned in drawing most of the outlines. The templates have been deliberately made so that the repositioning need not be done at the time the main part of the outline is drawn. Rather, it can be done anytime later. This is because the missing portion of the outlines is in most cases the straight, horizontal line that completes and closes in the bottom edge of the outline. Since this is easily visualized, the absence of this line normally does not impede the drawing of adjacent outlines.

With each opening in the template are printed alignment marks. These are made clear for ease of use. Since, typically, vertical centering is more critical than horizontal centering and is more difficult to achieve, the vertical alignment marks are longer than the horizontal ones.

The outlines drawn can conform in width, height, and configuration to the specifications established by the ANSI X3.5 Standard. It should be noted that these templates enable one to draw outlines of varying sizes that still conform to specifications of the Standard. The sizes have been chosen to permit an odd number of ten-to-the-inch (pica) typewritten letters on a line. Hence wording in the outlines can be easily centered if desired.

- Do-It-Now Exercise 2.2. Take your program template (P template) in hand or whatever template you have available. Orient it so that the printed words on the template are right-side-up. Then, keeping it in that orientation, find the rectangular opening in the template. (If you are using the P template, Figure 2-2 will help you.) Now, following the eight steps listed earlier in this chapter, draw one rectangle. If you are using the P template, draw the largest size. Then, at least four inches away in any direction, draw another rectangle. (Did you keep the top lines parallel to the top edge of your sheet of paper? You should have.)
- Do-It-Now Exercise 2.3. In the same manner, draw four rectangular outlines across the top part of a page and connect each to the adjacent one by a centered horizontal line. Then, below the first and last, draw two columns of two more rectangles, with connecting, vertical-centered lines. Then draw two more rectangles, horizontally placed and with centered connecting lines, to bridge the gap between the columns. (If you did your work neatly, you will now have a rectangle of rectangles. Are they uniformly spaced?

Are all the lines either horizontal or vertical? No tippy outlines, please, and "no fair" using layout lines.) If you are using the P template, repeat, using the smallest size rectangle.

- Do-It-Now Exercise 2.4. Repeat 2.3, using the parallelogram outline, the circle outline, and the diamond outline. Figure 2-3 shows the exercise done with the diamond outline. If you are using the P template, repeat, using the smallest size.

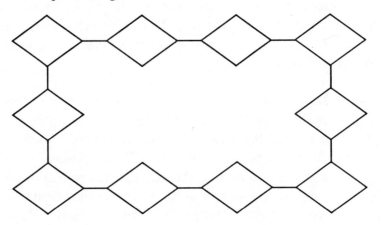

Figure 2-3. One solution to Do-It-Now Exercise 2.4

3

SYSTEM CHARTS

FUNCTION AND FORMAT

The system chart describes sequences of data-handling operations or processes which usually, but not necessarily, are done in part with an automatic computer. The system chart identifies the input data for each process, the output data from each process, and what the data-handling process is. The system chart tells what and how in broad terms. As indicated in Figure 2-1, the system chart does not tell the details of the process or the way in which the computer accomplishes a data-handling job. That is the work of the flow diagram, as is pointed out in the next chapter.

The basic format of the system chart follows a sandwich rule; that is, it is composed of alternating layers of data identifications and process identifications. The data identifications are equivalent to the bread of the sandwich, and the process identifications are equivalent to the filling in the sandwich. Just as sandwiches may be of the Dagwood type, so the output produced from one process operation may serve as the input for a following process operation (a compound system chart). But a system chart must always begin with inputs (data identifications) and must always end with outputs (data identifications).

This suggests that the outlines needed for preparing system charts are those for representing data, for representing processes, and for representing the sequence of processes and data identifications. The ANSI X3.5 Standard has outlines for each of these needs.

OUTLINES USED

Basic Outlines

The ANSI X3.5 Standard outlines are in three groups: the basic, the specialized, and the additional. Complete system charts can be prepared using only the basic outlines. The use of the specialized outlines is not necessary; but if they are used, they must be used in a consistent manner. The additional outlines are not normally needed in system charts.

For the outlines in each group, the shape is critical, but the size is not. This is because the Standard specifies the shape in two ways: (1) by the ratio of the width to the height, and (2) by the general geometric configuration. This means that a person preparing a system chart is free to draw outlines of any size to fit his own convenience. He may vary the size for the same outline type from place to place anywhere in his system chart, as long as he observes the ratio and general configuration specified. Use of a good template takes care of the shape problem and makes a selection of sizes easily available.

When one prepares a system chart, two points are worth attention from the outset. First, it is best to use a single width or weight of a line for drawing the outlines consistently throughout the entire system chart. Second, an outline has only one correct orientation. That orientation is illustrated in the figures in this book. As a general guide, portions of the outlines shown horizontally oriented are to be drawn that way.

The basic outlines specified are the input-output, the process, the flowline, and the annotation outlines. These are illustrated in Figure 3-1. All can be drawn by using either the P or S template.

The input-output outline indicates an input or output operation, or input or output data. It is defined for use irrespective of media, format, equipment, and timing. This outline can be drawn directly using the parallelogram opening in the S or P template. On the S template, it is in the far lower right; on the P template, it is in the far upper left and far lower right. Some specialized outlines may be substituted for this outline.

The process outline is the general-purpose outline. It is the de facto default outline for use when no other outline is appropriate. The process outline indicates data transformation, data movement, and logic operations. This outline can be drawn directly by using the rectangular opening in the S or P template. On both templates it is in the middle portion. Some specialized outlines may be substituted for this outline.

The flowline outline is a line or arrow of any length which connects successive other outlines to indicate the se-

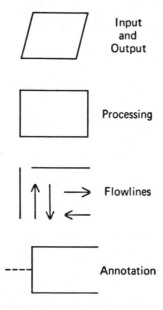

Input
and
Output

Processing

Flowlines

Annotation

Figure 3-1. Basic outlines for system charts

quence of operations on data. This sequence is termed the direction of flow. The flowline is for use in an alternating fashion with the other outlines. As such, it indicates the sequence in which the other outlines are to be read. To clarify the direction of flow or reading, open arrowheads may be used on any flowline as shown in Figure 3-1. This outline can be drawn by using any straight vertical or horizontal side of any opening or any edge of either template. Flowlines may be drawn with bends that are right angles. Other angles can be used, but are not recommended. For drawing the arrowheads, two thin diamond openings are provided in the center part of the S template. To use these, position the narrow tip of one of the thin diamonds on the "flow-going-out" end of a flowline, so that the flowline already drawn passes under the opposite tip of the same opening. Then trace along an equal distance of each edge of the thin diamond to draw a V-shape with its point on the end of the flowline.

The normal direction of flow is the normal direction of reading for people trained in the English language, from top to bottom, and from left to right. Where the flow follows this normal pattern, no open arrowheads are needed to remind the reader. In the event of any significant deviation from this pattern, arrowheads are required to signal the deviation to the reader's attention. Whenever the direction of flow might be ambiguous to a reader, arrowheads should be used to provide clarification. Bidirectional flow may be indicated by dual lines, each with open arrowheads, or, though this is less desirable, by open arrowheads in both directions on single flowlines.

The annotation outline as shown in Figure 3-1 provides a way to supply descriptive information, comments, and explanatory notes. The dashed line indicates that the line is not part of the line of flow and also serves as a pointer to the outline to which this annotation, explanation, or clarification applies. This outline can be drawn by using the rectangular opening in either the S or P template. Care is needed to remember to omit the right end of the rectangle, that is, to leave the rectangle open on the right. The dashed line, which is a part of the outline, may be connected to any of the three solid sides of the main part of the outline. It can be drawn in the same basic manner used for drawing flowlines.

Specialized Outlines

Groups. The specialized outlines fall into three groups. One group permits specification of the data-carrying media. (See Figure 3-2.) Another permits specification of the peripheral equipment type. (See Figure 3-3.) A third permits specification of selected types of processing action. (See Figure 3-5.) In each case, where no specialized outline has been provided, the basic outline covering the situation should be used. Thus, for any media or equipment, it should be the input-output outline. For any processing, it should be the process outline. The one exception is the communication link, for which the basic outline is the flowline. The openings needed to draw the specialized outlines are all on the S template.

Media Outlines. The document outline is the most commonly used of all the specialized media outlines. This outline, a stylization of a torn piece of paper, represents data in the form of hard copy input or output of any type. For example, it may represent data taking the form of printing on paper produced by a high-speed printer, or of marks on cards read by an optical reader, or of a graph produced by a data plotter, or of a page of typing produced on a terminal. This outline can be drawn directly by using the bib-shaped opening in the upper left part of the S template. Note that this outline is completed by drawing in the horizontal top line.

The magnetic tape outline is a circle with a horizontal, rightward-pointing line tangent to the bottom. This outline represents data in the medium of magnetic tape. It can be regarded as a specialization of the online-storage outline. The magnetic tape outline can be drawn by using any of the circle openings in the S or P templates, with an added horizontal line, one radius long, tangent to the right from the bottom of the circle.

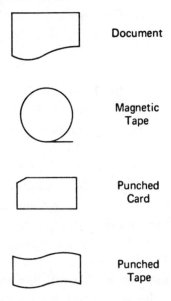

Document

Magnetic
Tape

Punched
Card

Punched
Tape

Figure 3-2. Specialized outlines for media for system charts

The punched-card outline represents data in the medium of a punched card of any style, size, or punching, such as Hollerith punched cards, binary punched cards, BCD punched cards, fifty-one column cards, and the like. Thus a time card which only has on it printed numbers and never is punched with equivalent or even unrelated information does not qualify for representation with the punched-card outline. (It takes a document outline instead.) Two further specialized forms of the punched-card outline are described later. The punched-card outline can be drawn directly by using the corner-cut, flat rectangular opening in the lower right part of the S template.

Equipment Outlines. The display outline is a stylization

of a CRT, with the face of the tube to the right and the neck of the tube to the left. This outline represents any kind of transitory data not in hard copy form, as for example CRT displays, console displays, and the like. This outline may also be used for intermediate output data used during the course of processing to control the processing. Common examples are the data produced on console printers and time-sharing terminals if the human user is expected to utilize immediately the data presented. Further, this outline has been used for CRT light-pen, cursor, joy-stick, or "mouse" input. But this is not a good use; the manual-input outline is a more realistic and appropriate choice. The display outline can be drawn in two steps from the bullet-shaped opening in the upper right part of the S template. To close the outline on the right-hand end requires repositioning the template. The closest curve to the right is for the smaller size; the tick marks on the other side of the curved opening mark off the closing arc for the small size. This outline is one of the harder ones to draw neatly because of its short curves.

The manual-input outline represents data acquired by human control of manually operated online equipment. Examples are data from the operation of keyboards, light-pens, console switch settings, push buttons, slides, transaction recorders, cursors, tag readers, and the like, where the human operator provides the timing. This outline is a stylization of a side view of a keyboard. The outline can be drawn directly by using the sloping-top trapezoid in the center right part of the S template.

The communication-link outline is represented by a zig-zag flow line. This is appropriate because, typically, data communication done with the aid of equipment provides a flow of data from one place, or from one medium or equipment, to another. As such, even though the outline is equipment-oriented, it is used as a specialized flowline. Where necessary, open arrowheads may be used to indicate the direction of flow, in the manner previously described. The

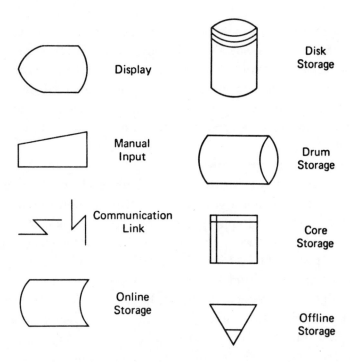

Figure 3-3. Specialized outlines for equipment for system charts

outline can be drawn directly by using the Z-shaped opening at the far upper left of the S template. The opening makes possible zigzags of two different sizes. The template may be turned to make the zigzag go in any desired direction, like a flowline.

The online-storage outline represents data held in any online-intermediate and online-external-storage device of any type, as, for example, magnetic disks, magnetic tapes, magnetic drums, magnetic cards, semiconductor auxiliary storage, additional banks of magnetic core storage, microfilm, etc. For data on some of these devices, more specialized outlines are available when more precise specification is desired. A more specialized outline for data on magnetic tape has already been noted. The online-storage outline can be drawn by using the log-shaped opening in the center part of the S template. To close the right end, use the same procedure employed for the display outline. The rightmost curve is for the smaller outline.

The disk-storage outline represents data stored on a disk device of any type, especially a magnetic disk. The outline is a stylization of a cylinder standing on end. The outline is a further specialization of the online-storage outline. This outline requires turning the S template one quarter of the way around (clockwise is best), then using the log-shaped opening. But, completing the outline requires four additional lines. These are best put in by repositioning the template to draw in the four pendant curves by using the convex opening now just below the bullet-shaped opening. Since this is the most difficult outline to draw, the main stages are shown in Figure 3-4.

Figure 3-4. Drawing the disk outline with the S template

The drum storage outline represents data stored on a drum device, especially a magnetic drum. The outline is a stylization of a drum lying on its side. The outline is a further specialization of the online-storage outline. This outline too can be drawn using the log-shaped opening. The usual procedure is to draw the online-storage outline and then, using the convex opening to the right of the bullet-shaped opening, complete the right end of the drum outline.

The core-storage outline represents data stored in a magnetic core, semiconductor, or similar high-speed storage device that is *not* the primary internal storage for the computer. It might be, for example, an auxiliary, online, bulk, magnetic-core device, or a remote computer which is connected online with the computer doing the main processing. The outline is a stylization of two drive lines in a magnetic core array. This outline too is a further specialization of the online-storage outline. The core-storage outline can be drawn by using the square opening at the top center of the S template. This must be repositioned twice to get the two inner lines.

The offline-storage outline is an equilateral triangle with a small bar. This outline represents any data stored offline regardless of the medium and regardless of the equipment used. In common practice, it is used for manually maintained data files. This outline can be drawn by using the downward-pointing triangle in the lower left part of the S template. The tick mark indicates the approximate position for the crossbar. This can be easily put in at the same time as the line closing in the top of the outline.

Process Outlines. Seven specialized process outlines find use in system charts. These are the manual, auxiliary, merge, extract, sort, and collate operation outlines and also the pre-defined-process outline. All except the manual and auxiliary outlines may indicate processes performed either within, or external to, an automatic computer.

The manual-operation outline indicates any offline-output- or offline-input-producing operation which has its speed determined by the speed of the human operator. Examples are entering data offline by means of a keyboard as in a keyboard-to-magnetic-tape operation, and finding a folder in a file cabinet drawer. This outline can be drawn directly by using the trapezoidal opening in the lower left corner of the S template.

The auxiliary-operation outline indicates any offline operation performed on equipment which operates at its own speed or at a speed determined by something other than the speed of its human operator. Examples of auxiliary operations are card-sorting operations, punched-card-interpreting operations, and the like. Auxiliary operations are performed predominantly by equipment, not by human beings. This auxiliary-operation outline can be drawn directly by using the square opening in the top center of the S template.

The merge outline indicates the creation of one set of items from two or more sets having the same sort sequences. This outline requires multiple, incoming flowlines, but one outgoing flowline only. The outline may be used for both online and offline operations and also for both computer-performed and noncomputer-performed merges. The merge outline can be drawn directly by using the downward-pointing, triangular opening in the lower left part of the S template.

The extract outline indicates the reverse of the merge; that is, it indicates the creation of two or more sets of items from, and in, the same sort sequence as the original set. Hence it requires an incoming flowline and multiple, outgoing flowline. The outline may be used for both online and offline operations and also for both computer-performed and non-computer-performed extracts. The extract outline can be drawn directly by using the upward-pointing, triangular opening in the lower left center part of the S template.

The sort outline indicates the sorting of a set of items into some sequence on the basis of some (usually specified) key. The outline may be used for both online and offline operations and also for both computer-performed and non-computer-performed sorts. The sort outline can be drawn by using first the upper, and then the lower, triangular outlines on the S template.

The collate outline indicates a combination of merge and extract. Thus this outline requires more than one incoming (entrance) flowline and more than one outgoing (exit) flowline. This definition of collate is not fully consistent with the usual definition of collate, but is used in the ANSI Standard.[1]

---

[1] See the entry for "collate" in, for example, Martin W. Weik, *Standard Dictionary of Computers and Information Processing*, New York: Hayden Book Company, Inc., 1969.

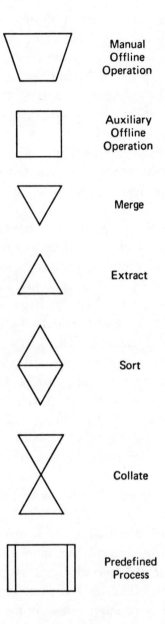

Figure 3-5. Specialized outlines for processes for system charts

The outline may be used for both online and offline operations and also for both computer-performed and noncomputer-performed operations. The collate outline can be drawn by using first the lower, and then the upper, triangular outlines on the S template.

The predefined process outline indicates or identifies one or more operations which are specified in more detail elsewhere, as in a booklet, or in a different system chart (but not in another part of this same system chart). This outline is rarely used in system charts. This is probably because such outside processes at the system level are usually complex and produce many outputs. The predefined process outline can be drawn by using the rectangular opening in the center part of the S template. The template must be repositioned in order to draw the two vertical inner lines.

Additional Outlines

Two of the additional outlines find some use in system charts: the parallel-mode and the connector outlines. As noted later, the use of the connector is best avoided in system charts. These outlines are shown in Figure 3-6.

The parallel-mode outline is a pair of horizontal lines

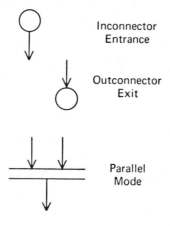

Inconnector
Entrance

Outconnector
Exit

Parallel
Mode

Figure 3-6. Additional outlines for system charts

with one or more vertical-entry flowlines and one or more vertical-exit flowlines. It is used to indicate the start or end of simultaneous operations. It can be drawn with any straight edges of either the S or P template. If this outline is the first in a series of outlines, it has no entrance flowlines; if it is the last, it has no exit flowlines.

The connector outline, a circle, is best not used in system charts. If used, it must be used at least in pairs. To that end, two forms of the connection can be distinguished, based upon the flowlines associated: the inconnector or entry connector and the outconnector or exit connector. An inconnector or entrance has a flowline leaving it, but none entering it; an outconnector or exit has a flowline entering it, but none leaving it. Each inconnector may have from zero through any number of outconnectors associated with it. However, each outconnector must have exactly one inconnector associated with it. One function of the connector outline is to enable a long sequence of outlines (a *flow*) to be broken into pieces to fit conveniently on a page. The connector outline also provides ways of joining together convergent lines of flow that fan-in to some particular point. In addition, it provides a way of identifying divergent lines of flow (fan-out). The connector outline can be drawn by using any circle opening of either the S or P template.

SYSTEM-CHART CONVENTIONS

Cross-References

In order to make cross-reference easy between parts of the system chart, two conventions are common. One is to use or to assign names to portions of the flow represented by the flowchart. These names often are the names used to designate parts of the system.

An alternative convention (not mutually exclusive of the other) is to identify a location on each physical piece of the system chart, as, for example, in terms of page, row, and column, as in the manner of map coordinates. An example of such a coordinate scheme is given in Figure 3-7. Another scheme that is also common is to assign a sequential number to each outline on a page following the line of flow, and then to record this number beside each outline.

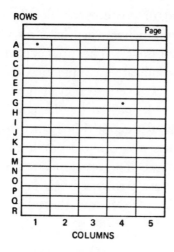

ROWS

*Example: thus, 9A1 is on page 9, row A, column 1; and 9G4 is on page 9, row G, column 4

Figure 3-7. One coordinate-location, cross-reference scheme for system charts

The ANSI Standard is in conflict with the ISO Standard and with previous usage in the United States on the handling of references. The ISO and common American usage has been to place the identifying name immediately above and to the left of an outline and to place the location reference above and immediately to the right of an outline. The ANSI Standard advances exactly the opposite convention, but recognizes and cites the deviation from ISO Standard.

In this book the ISO convention is used since it is also commonly used in the United States and since the ANSI Standard explicitly recognizes the ISO position.

Crossing Flowlines

The Standard makes specific provision for connectors and cross-references. These can be used to avoid the necessity of using crossing flowlines.

If it is desired to use crossing flowlines, then the convention is that the flowlines shall have no arrowheads in the vicinity of the crossover. The presence of arrowheads on the flowlines is to indicate a conjunction or coming together (fan-in) of the flow. The direction of flow at such points of joining flowlines is designated by the position and direction of the arrowheads, as exampled in Figure 3-8.

In system charts it is often desirable to show the fan-in and fan-out of flows. The usual practice for making these unambiguous is to use arrowheads sparingly and to place the flowlines consistently to follow the usual pattern of flow,

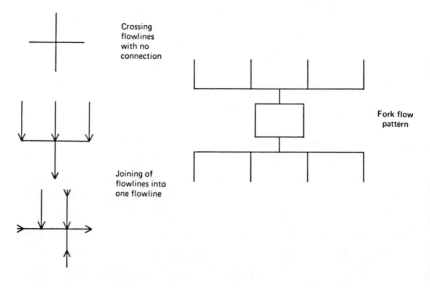

Figure 3-8. Conventions for flowlines for system charts

from top to bottom, and from left to right. Very common patterns, for which the arrowheads are often omitted, are shaped like forks. If the lines are downward or to the right, the pattern represents fan-out; if the lines are upward or to the left, it represents fan-in. These patterns are illustrated in Figure 3-11.

Multiple Outlines

Because the situation is frequently encountered, conventions exist for the representations of multiple instances of the specialized media outlines. These take two forms: one specifically for punched-card media and the other for media generally (which can also be used for punched cards). The one applicable to punched cards provides only for representing a deck of cards and a card file, as shown in Figure 3-9. These find their most common use in system charts for systems implemented with punched-card handling equipment.

The more general convention is for use when multiple forms of specified media have different identifications and uses. For example, in punched-card installations, it is common to have a master, or header, card followed by several detail, or trailer, cards. The convention for representing this situation is that the main or first medium outline should be drawn in full. Following it and partially obscured by it in any clockwise position from it, a partial but closed outline of other instances of the same medium may be drawn in sequence, as shown in Figure 3-9.

With this convention, flowlines need special attention. One convention is that flowlines may enter or leave from any part of the multiple-outline group. This, in effect, treats the group as though it were a single outline that simply has inner lines marking off interior portions.

But another convention is needed for the common case where the group is to be broken into component parts and the component parts processed differently. The convention

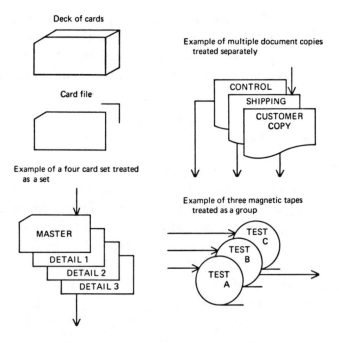

Figure 3-9. Multiple media conventions

used here requires that if a single flowline enters and a single flowline exits, then the multiple media shown are treated as a group. If multiple flowlines are used on either the entrance or exit sides, then the multiple flowlines apply only to the specific media outlines to which they are connected individually.

Basic Rule

The basic rule in the drawing and the interpretation of system charts is the sandwich rule, as noted earlier. This rule may be applied in simple form, where the input and output data are the equivalent of the bread, and the processing is the equivalent of the filling. Or the rule may be applied in compound form (like a Dagwood sandwich) with interior alternating layers of data (bread) that are the output of one process (filling) and the input for another (filling).

To see this sandwich rule in use, consider the creation of a system chart using only the basic outlines. Assume that the input available is a set of data about the ages of employees. Assume that the output desired is a single number, the average age of employees. If the basic outlines are used, no attention need be paid to the media or to the equipment. Hence, as shown in Figure 3-10, the system chart begins with an input-output outline for the input. Connected to that by a flowline is a process outline. Connected to that by a flowline and ending the system chart is an input-output outline for the output.

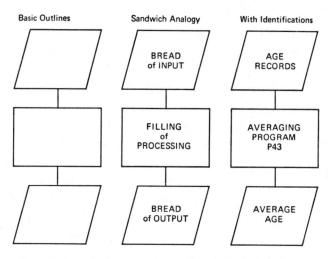

Figure 3-10. A simple system chart using only the basic outlines

To summarize, the basic format of the system chart is a sandwich. It always begins with the bread of input and ends with the bread of output. The sandwich filling is the processing, which converts the input into the output. But the system chart does not tell how the processing is accomplished; it only identifies the processing done.

Identifications

The bare outlines shown in Figure 3-10 are meaningful to someone who has clearly in mind the identification of the input, the output, and the processing. But it has less communication value to others because, even though it tells that input is to be converted into output, it does not identify which particular input, or what particular processing, or which particular output is meant. A common convention, therefore, for improving the communication value of the system chart is to indicate, within the outlines, the identification of each input, each output, and each process.

For this purpose, the usual convention is to use the names normally assigned at an installation to the input and output. If the system chart is likely to be read by persons not conversant with those names, then the English language equivalent may be written out in full within the outlines to provide the identification. Thus Figure 3-10 also provides a restatement that incorporates the identifications absent from the system chart on the left in Figure 3-10.

This same system chart is also present in the top part of Figure 3-11. But Figure 3-11 also illustrates more since it shows a compound system chart rather than just a simple one. The average which was the output of the first processing operation, in turn, serves as input for another operation. Here the average age of employees is to be combined with previously calculated data on the average age of employees to produce a chart showing the trend of the average age of employees over the course of time. The preparation of this chart is a separate processing operation from a computation of the current average age. The other output of the trend program is the updated record of the prior averages.

Here, as before, clear identification is provided of each of the inputs, output, and processings, but no indications of the nature of the medium or equipment are provided in the

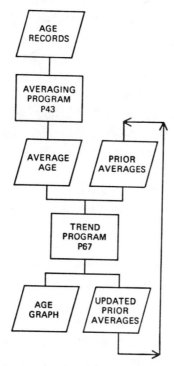

Figure 3-11. Compound system chart using only the basic outlines

choice of outlines. Note the compound (Dagwood sandwich) structure.

- Do-It-Now Exercise 3.1. Describe in words the system shown in Figure 3-11. What information would you have to add to what you can infer from the system chart in order to give a more complete description of the system? How much of that information might be gleaned from flow diagrams? From where might the remaining data come? (Hint: Try asking such questions as: what, how, when, who, where, and how much.)
- Do-It-Now Exercise 3.2. Using only the basic out-

lines, prepare a system chart for the following system: Data about the performance of a space probe must be converted from a gray code to a floating-point, hexadecimal code. Then these data must be printed and, at the same time, inserted in a set of similar data from other probes.

- Do-It-Now Exercise 3.3. Using only the basic outlines, prepare a system chart for the following system: An executive's report request must be translated into control cards for a report-generator run. That run uses data from a sales-data file, and as it produces the report, writes the summary data back into the file.

The system-chart creator has many options for stating the identifications. The practice illustrated in this chapter is to use upper case (capital) letters for the names of specific input and output data and of specific processes. Some people also put annotation and all identifications, whether specific or not, in upper case letters. This practice is illustrated later in this chapter; an alternative practice is illustrated in Chapter 5.

Specialized Outlines

In order to improve the communication value of the system chart still more, specialized outlines may be used in place of the basic outlines already presented. Thus, Figure 3-12 presents a redrawing of Figure 3-11 that uses the specialized outlines. Figure 3-12 shows that the data on the age of employees is on cards and that the output of the average age is put onto a magnetic disk or other external storage device, where it serves as input to the trend program.

The other input to this trend program is from a magnetic tape, which has recorded on it the previously computed average ages. A magnetic tape also receives the output from

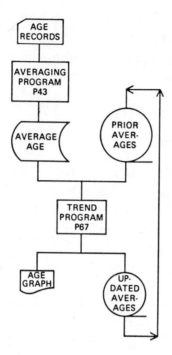

Figure 3-12. Compound system chart using the specialized outlines

this program so that it can be recycled to serve again as input if necessary.

The hard copy output from the second program is a time-series graphic plot. Assuming that the stress is on the hard copy aspects of this output, then the document outline is the appropriate specialized outline to use. If this plot, however, were displayed on a CRT, then a display outline would be the appropriate choice.

- Do-It-Now Exercise 3.4. Using the specialized outlines, redo Do-It-Now Exercise 3.2, assuming that the raw probe data are on magnetic tape, the floating-point data are on magnetic drums, the prior data are on magnetic tape, and the printed data are on paper. Assume that both operations are computer runs.

- Do-It-Now Exercise 3.5. Using the specialized outlines, redo Do-It-Now Exercise 3.3, assuming that the
typewritten request is translated by a programer, the
report is on a CRT, and the sales-data file is on a
magnetic disk.

Use of Connectors

If a compound system chart requires more space on the
page to represent it than is available, then connectors may
be used to break the chart into parts and to indicate the connection between the parts on the separate pages. This procedure is used by some analysts, but does not improve the
communication value of the system chart. An alternative
procedure gives superior results.

The matter may again be likened to the Dagwood sandwich. If one makes a Dagwood sandwich that is too large to
bite, one procedure is to break the sandwich into two or
more parts. Whenever doing this, however, one must add an
additional slice of bread for each break because each complete sandwich must begin with a slice of bread and end
with a slice of bread. Since, by analogy, input or output data
serve as slices of bread, one may break a long system chart
into shorter charts and still show a connection between
them by the choice of representing input or output data. To
see this, imagine breaking the system chart in Figure 3-12
into two separate system charts.

To make the communication value of the system chart
high, one should show all of the inputs and outputs for each
part of the system. If one of these appears at one place as an
output, and at another as an input, then one has only to
repeat the outline with the identical identification and provide
an appropriate cross-reference. This is illustrated in Figure
3-13. Notice that the disk output from the first run, which
serves as an input to the second run, is shown twice. This is
considerably more illuminating to someone who studies any

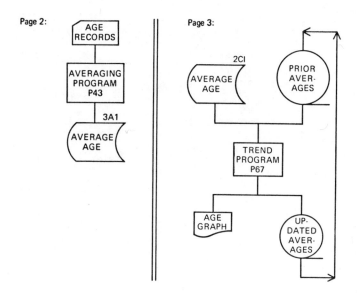

Figure 3-13. Example of good practice in breaking a system chart

part of the chart than would be the alternative of using a connector outline, illustrated in Figure 3-14, where two different ways of doing it are shown.

In summary, the most effective way to break a system chart into parts in order to fit it onto pieces of paper of limited size, is to repeat the representation for selected inputs or outputs, identifying and cross-referencing them appropriately. In this way, the material shown on each page is complete in itself.

Annotation

When one breaks a system chart, a problem arises in connection with identifying the source and use of data. The clearest convention is, as shown in Figure 3-13, to use cross-references with an exact repetition of the data identification. This also serves well for multiple uses of an output as inputs to several process outlines.

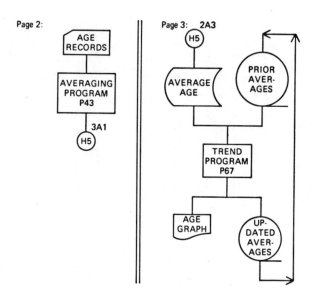

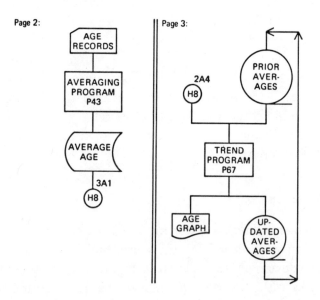

Figure 3-14. Examples of poor practice in breaking a system chart

One alternative way of handling the situation is by the use of the wording within the outlines themselves. But this clutters the space conventionally used for identification, making it a dual-use space, not a single use. This decreases the communication value of the chart, and it is not illustrated here.

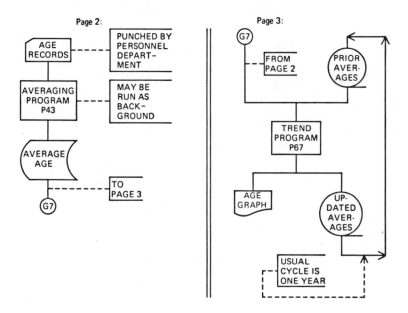

Figure 3-15. Annotation outlines in system charts

Another way of maintaining the identification of the source and use of data is by the use of an annotation outline. For example, one could be inserted at the bottom of the first page, indicating the page to look on to find the additional use of this output as an input, as shown in Figure 3-15. On the page on which it appears as an input, the annotation outline could again be shown with an indication in it as to where this input came from as an output. It is clear from even a casual examination of Figure 3-15 that the excessive use of annotation outlines clutters the system chart and can decrease its communication value.

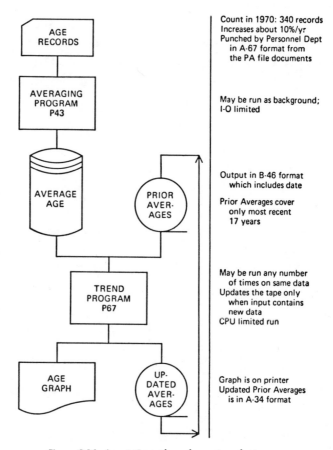

Count in 1970: 340 records
Increases about 10%/yr
Punched by Personnel Dept
in A-67 format from
the PA file documents

May be run as background;
I-O limited

Output in B-46 format
which includes date

Prior Averages cover
only most recent
17 years

May be run any number
of times on same data
Updates the tape only
when input contains
new data
CPU limited run

Graph is on printer
Updated Prior Averages
is in A-34 format

Figure 3-16. Annotation column for system charts

If one is to avoid the clutter of the annotation outlines and yet provide the annotation, an alternative is to add a column of annotation to the right (or the left) of the system chart. This can provide information that is frequently helpful in interpreting system charts, as exampled in Figure 3-16. The information most commonly shown in such annotation columns is the volume of the input and output, the timing of the availability of input or output, the control procedures, the equipment configurations required, the personnel complements, and the geographic locations. These items, as a rule, are not conveniently shown in the system chart itself because their presence diverts attention from the main flow; they are not of concern to all readers. The annotation column is thus an augmentation of the ANSI Standard.

GUIDELINES FOR SYSTEM CHARTS

For preparing system charts the following guidelines have been found helpful from experience. The first guideline is to choose the wording within the outlines (the data and process identification names) to fit the needs of the readers of the system chart. A chart using something approaching the English language, as illustrated in Figure 3-16, can be widely understood. A chart which uses only specialized names can be fully understood only by those who know the specialized names. Thus, the system chart shown in Figure 3-18 is less easily understood than the same system chart shown in Figure 3-16.

The second guideline is to use the data and process identification names consistently and to keep them brief or short. (Compare Figures 3-16 and 3-18.) If the same name appears more than once anywhere in the system chart, it should always identify the same thing.

The third guideline is to use as small a size as possible for the outlines. This improves the communication value of the chart because it makes possible a more compact layout. It assists the reader's eye to take in more at one glance.

The fourth guideline is to leave blank space around major nonconvergent flows. This sets them off visually and makes their role in the system more easily comprehended, as in Figure 3-17. By contrast, in a uniformly packed or tightly spaced system chart, even simple straight lines of flow are difficult to see clearly.

The fifth guideline is to collect incoming flowlines and

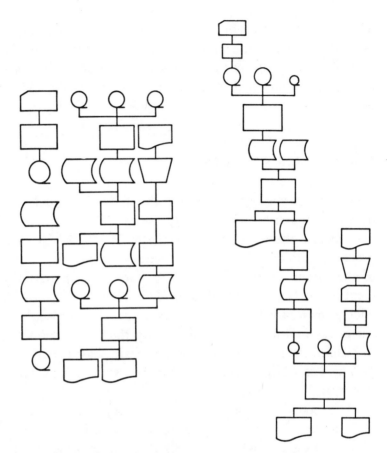

Figure 3-17. Example of an open and a tight layout of a system chart
for the same system

outgoing flowlines so that the flowlines actually entering and leaving a processing outline are kept to the minimum. This is illustrated in Figure 3-16.

The sixth guideline is to minimize crossing flowlines. Crossing flowlines can be eliminated by repeating input or output outlines with appropriate cross-reference as to source or destination.

The seventh guideline is to use the specialized outlines wherever possible. (Compare Figures 3-11 and 3-12.) Their use improves the communication value of the system chart. They are only slightly more difficult to draw.

The eighth guideline is to be consistent in the use of the specialized outlines. The specialized process outlines are not very much of a problem. However, for the equipment and media outlines for data, three levels are possible. Thus data on magnetic disks can be represented by (1) the basic input-output outline, or (2) the online-storage outline, or (3) the magnetic disk outline. The person preparing a system chart should pick a level and then adhere to it consistently and not jump around in using the levels of outline.

The ninth guideline is to use cross-reference and annotation generously, but not to excess. The more the system chart can tell the reader quickly and easily, the more valuable it is. Ways of providing cross-reference and annotation have been noted previously.

The tenth guideline is to give particular attention to the processing that affects data prior to the time that the data become input to a computer program or run. These are most often manual operations and auxiliary operations. Failure to specify them in full is one of the most common shortcomings of system charts. (Example: How was the data punched into the cards in the system shown in Figure 3-18?)

The eleventh guideline is to use the ANSI Standard and

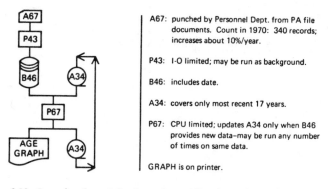

A67: punched by Personnel Dept. from PA file documents. Count in 1970: 340 records; increases about 10%/year.

P43: I-O limited; may be run as background.

B46: includes date.

A34: covers only most recent 17 years.

P67: CPU limited; updates A34 only when B46 provides new data–may be run any number of times on same data.

GRAPH is on printer.

Figure 3-18. Example of specialized wording within the outlines of a system chart

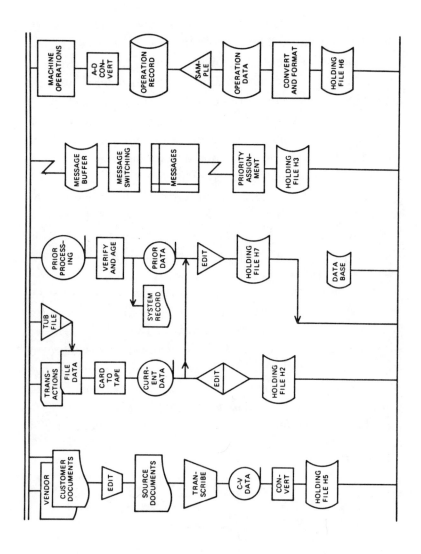

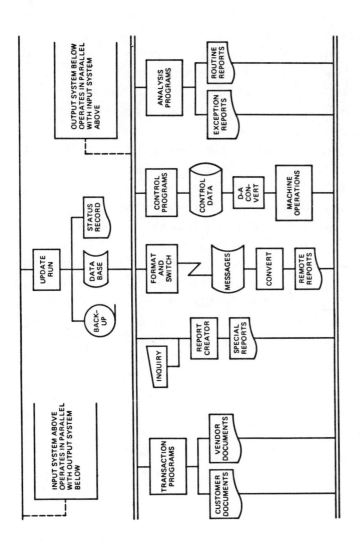

Figure 3-19. A portion of an extensive system chart

to shun deviations. A few examples may clarify this. (The distinction between a violation, a deviation, and an augmentation was noted earlier.) The use of closed arrowheads on flowlines is an example of a violation. Another example is using a circle with a short right-of-center horizontal line touching the bottom center as an outline for data on punched cards. This is a violation because the ANSI Standard assigns a specific significance to such a decorated circle and provides a specific outline for data on punched cards.

Three examples of deviations from the ANSI Standard are provided by IBM's flowchart template.[2] IBM advances an outline for *keying*. This is a serious deviation, since the ANSI Standard already provides two outlines for data from key-driven equipment or for such equipment's operation. These are the manual input (online) and the manual operation (offline). IBM advances an outline for a *transmittal tape*. This is a kind of document. The Standard provides an outline for documents generally. IBM advances an outline for an *offpage connector*. The Standard provides a connector outline and specifies the use of cross-reference to indicate the location.

An example of an augmentation or elaboration of the ANSI Standard would be the use of a five-pointed-star outline to represent data acquired by the online operation of an optical detector of particle tracks in a spark chamber. Such equipment is not generally available, and hence the ANSI Standard provides no specialized outline. The basic input-output outline is applicable, and thus no specialized outline is really necessary.

[2] IBM Corporation, *Flowcharting Template X20-8020*, New York: IBM Corp., 1969, one plastic cutout drawing guide in a printed envelope.

4

FLOW DIAGRAMS

FUNCTION AND FORMAT

The flow diagram describes sequences of data-handling operations performed internally within an automatic computer. The flow diagram names the input data and designates in sequence the operations the computer applies to the data. The flow diagram tells in specific terms what operations the computer uses on which data, and in what sequence. As indicated in Figure 1-1, the flow diagram concentrates on describing the process used to transform input data into output.

The flow diagram is algorithm-oriented. As such, its primary emphasis is on depicting the sequence of operations that tell how the computer transforms data. The secondary emphasis is on identifying the portions of the data structures affected and the operations performed. Questions of media or equipment typically become trivial.

Since the operations to transform data structures commonly consist of long sequences of actions, the character of the flow diagram differs considerably from that of the system chart. In the system chart, a sandwich rule describes its basic structure. No such convenient rule serves in the case of the flow diagram. Since the flow diagram is a detailed elaboration and extension of what is usually shown as a single outline on a system chart, it requires many more outlines and a much more extensive presentation of specifics than does the system chart.

The general character of a flow diagram, therefore, is that of a sequence of alternating flowline and process outlines. Somewhere in the early portion of this sequence will usually appear one or more input-output outlines to indicate the input of a data structure. Near the end of the sequence will usually appear one or more input-output outlines to indicate the output of a data structure.

Because of their greater length, the flow diagrams must be broken into parts as a practical matter. For this reason, connector usage and cross-referencing become important considerations in the creating and reading of flow diagrams. The outlines advanced by the ANSI X3.5 Standard cover those needed for flow diagrams.

OUTLINES USED

Basic Outlines

The ANSI X3.5 Standard outlines are in three groups—the basic, the specialized, and the additional. Complete flow diagrams can be prepared by using the basic and additional outlines. The use of the specialized outlines is not necessary; but if they are used, they must be used in a consistent manner to present the processes adequately.

For the outlines in each group, the shape is critical, but the size is not. This is because the Standard specifies the shape in two ways: (1) by the ratio of the width to the height, and (2) by the general geometric configuration. This means that a person preparing a flow diagram is free to draw outlines of any size to fit his own convenience. He may vary the size for the same outline types from place to place anywhere in his flow diagram as long as he observes the ratio and general configuration specified. Use of a good template takes care of the shape problem and makes a selection of sizes easily available.

When one prepares a flow diagram, two points are worthy of attention from the outset. First, it is best to use a single width or weight of line for drawing the outlines consistently throughout the entire flow diagram. Second, an outline has only one correct orientation. That orientation is illustrated in the figures in this book. As a general guide, portions of the outlines shown horizontally oriented are to be drawn that way.

The basic outlines specified are the input-output, the process, the flowline, and the annotation outlines. These are illustrated in Figure 4-1. All can be drawn by using either the P or S template.

The input-output outline indicates an input or output operation, or input or output data. It is defined for use irrespective of media, format, equipment, and timing. This outline can be drawn directly by using the parallelogram opening in the P or S template. On the P template it is in the far upper left and far lower right; on the S template it is in the far lower right. Some specialized outlines may be substituted for this outline.

The process outline is the general-purpose outline. It is the de facto default outline for use when no other outline is appropriate, and it is by far the most common, nonflowline outline in most flow diagrams. The process outline indicates data transformation, data movement, and logic operations. This outline can be drawn directly by using the rectangular opening in the P or S template. On both templates it is in the middle portion. Some specialized outlines may be substituted for this outline.

The flowline outline is a line or arrow of any length which connects successive other outlines to indicate the sequence of operations or data. This sequence is termed the direction of flow. The flowline is for use in an alternating fashion with the other outlines. As such, it indicates the

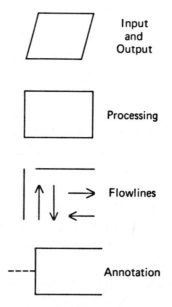

Input
and
Output

Processing

Flowlines

Annotation

Figure 4-1. Basic outlines for flow diagrams

sequence in which the other outlines are to be read. To clarify the direction of flow or reading, open arrowheads may be used on any flowline, as shown in Figure 4-1. This outline can be drawn by using any straight vertical or horizontal side of any opening or any edge of either template. Flowlines may be drawn with bends that are right angles. Other angles can be used, but are not recommended. For drawing the arrowheads, two thin diamond openings are provided in the left part of the P template. To use these, position the narrow tip of one of the thin diamonds on the "flow-going-out" end of a flowline, so that the flowline already drawn passes under the opposite tip of the same opening. Then trace along an equal distance of each edge of the thin diamond to draw a V-shape with its point on the end of the flowline.

The normal direction of flow is the normal direction of reading for people trained in the English language: from top to bottom and from left to right. Where the flow follows this normal pattern, no open arrowheads are needed to remind the reader. In the event of any significant deviation from this pattern, arrowheads are required to signal the deviation to the reader's attention. Whenever the direction of flow might be ambiguous to a reader, arrowheads should be used to provide clarification. Bidirectional flow may be indicated by dual lines, each with open arrowheads, or, though this is less desirable, by open arrowheads in both directions on single flowlines.

The annotation outline as shown in Figure 4-1 provides a way to supply descriptive information, comments, and explanatory notes. Its dashed line indicates that it is not part of the line of flow and serves as a pointer to the outline to which this annotation (explanation or clarification) applies. This outline can be drawn by using the rectangular opening in either the P or S template. Care is needed to remember to omit the right end of the rectangle, that is, to leave the rectangle open on the right. The dashed line, which is a part of the outline, may be connected to any of the three solid sides of the main part of the outline. It can be drawn in the same basic manner used for drawing flowlines.

Additional Outlines

The additional outlines are for the convenience of the reader and not for the purpose of describing data-processing action. These outlines provide for handling the limitations of pages of various sizes and make it more convenient to show connections in the sequences of flow. These outlines are shown in Figure 4-2.

The connector outline, a circle, must in practice be used at least in pairs. To that end, two forms of the connection can be distinguished, based on the flowlines associated: the inconnector or entry connector and the outconnector or exit connector. An inconnector, or entrance, has a flowline leaving it, but none entering it; an outconnector or exit has a flowline entering it, but none leaving it. Each inconnector may have from zero through any number of outconnectors associated with it. However, each outconnector must have exactly one inconnector associated with it. One function of the connector outline is to enable a long sequence of outlines (a *flow*) to be broken into pieces to fit conveniently on a page. The connector outline also provides ways of joining together convergent lines of flow that fan-in to some particular point. In addition, it provides a way of identifying divergent lines of flow (fan-out). The connector outline can be drawn by using any circle opening of either the P or S template.

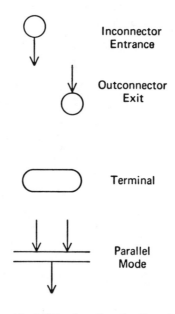

Figure 4-2. Additional outlines for flow diagrams

The terminal-connector outline serves to indicate a beginning, an end, or a break in the usual line of flow. In the first two uses, it substitutes for an ordinary connector at the beginning and the end of major portions of a sequence of outlines (a *flow*), particularly when these portions are identified by a name, as, for example, for a closed subroutine. In its third use, it may represent a start, stop, halt, delay, pause, interrupt, or the like. For this use it has both an entry and exit flowline. This outline can be drawn directly by using the ellipse opening in the bottom center of the P template. Note that only two sizes are provided.

The parallel-mode outline is a pair of horizontal lines with one or more vertical-entry flowlines and one or more vertical-exit flowlines. It is used to indicate the start or end of simultaneous operations. It can be drawn with any straight edges of either the P or S template.

Specialized Outlines

Groups. The specialized outlines fall into three groups. One group permits specification of selected types of processing action. (See Figure 4-3.) The other two permit specification of the data-carrying media and of the peripheral equipment type.

The specialized media and equipment outlines are not illustrated here, for they have little place in flow diagrams. One reason for this is that in a flow diagram the emphasis is on the data transformations accomplished by the arithmetic and logic unit, not on the equipment or media used to carry data. A second reason is that, with modern operating systems, the management of a computer facility at run time may usually change the media or equipment assigned to provide input or receive output data. The few instances of equipment- or media-dependent operations (as in some communications work) deserve the use of annotation outlines to call special attention to their particular significance. Certainly the flow diagrams for almost all FORTRAN, COBOL, ALGOL, PL/1, Autocoder, and BASIC programs can be spared the use of the specialized media and equipment outlines. Only for system software routines designed to substitute for the I-O routines provided by the operating system, can a good case

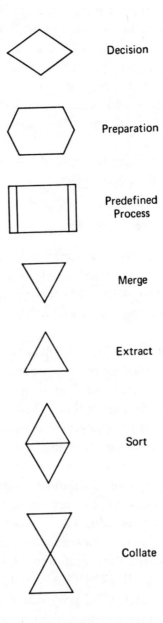

Decision

Preparation

Predefined
Process

Merge

Extract

Sort

Collate

Figure 4-3. Specialized outlines for processes for flow diagrams

be easily made for the use of the specialized media and equipment outlines.

As in the case of system charts, where no specialized outline has been provided, the basic outline covering the situation should be used. Thus for any media or equipment, it should be the input-output outline. For any processing, it should be the process outline. The openings needed to draw the specialized outlines needed for flow diagrams are all on the P template.

Process Outlines. The most common of the specialized process outlines for flow diagrams is the decision outline, shown in Figure 4-3. It indicates comparison, decision, testing, or switching operations, which determine or select among a variety of alternative flows (sequences of operations). As such, the number of flowlines leaving a decision outline must always be greater than one. The decision outline can be drawn using the half-diamond openings in the left part of the P template. The usual practice is to draw the top half first, and then the bottom half, after repositioning the template to match the corners. Typically, programers and analysts complain that the diamond outlines are too small. For this reason, the P template provides for drawing decision outlines of a generous dimension.

Sometimes the number of flowlines leaving a decision outline (the number of exits) exceeds three. In this case, the ANSI Standard advances, as shown in Figure 4-4, several alternatives considered to be equally acceptable. One is the organizational chart tree-pattern of flowlines from a single flowline leaving a decision outline. Another alternative used in the same way is a vertical flowline which has a number of horizontal flowlines from it. To save, one may omit formal outconnectors from these two forms. Quite a different alternative is a "branching" table in the form of a series of pairs of rectangles, packed together in a double row or a double column. The upper or left portion replaces the decision outline; the lower or right portion replaces the usual outconnectors. The tables use the first of each of the pairs of boxes to explain the control branching, and the second to indicate the place.

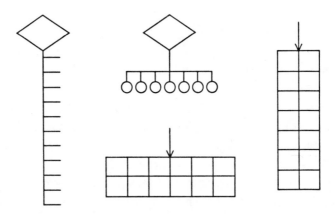

Figure 4-4. Outlines for large numbers of decisions in flow diagrams

The preparation outline indicates operations on the program itself. They are usually control, initialization, cleanup, or overhead operations not concerned directly with producing the output data, but usually necessary to have done. Three examples are setting the limiting value as an iteration control, decrementing an index register, and setting the value of a program switch. By convention, when either a decision or a preparation outline could be equally well used (such as for testing a switch), the common practice is to use the decision outline. The preparation outline can be drawn directly by using the pointed-end openings in the right-hand part of the P template.

The preparation outline is sometimes more widely used than just for operations on the program data structures. Some programers use it for all "red-tape" or "housekeeping" operations, such as the passing of parameters in the calling of routines, the initialization of working storage areas, and the clearing of output buffer areas. Clearly the dividing line is hazy between this extended "program modification" or "red tape" view, and the program data structures view. It can be argued, for example, that the data used temporarily to hold places in a buffer area are really part of the program data structure. In the same view, it can be argued that a literal imbedded in an instruction is part of the operand data structure.

The predefined-process outline indicates or identifies one or more operations which are specified in more detail elsewhere, as in a booklet, or in a different flow diagram (but not in another part of this same flow diagram). Examples of a predefined-process are a named, closed subroutine or a routine from the operating system for the computer. The predefined-process outline can be drawn by using the rectangular opening in the center part of the P template. The template must be repositioned in order to draw the two vertical, inner lines.

The other specialized process outlines sometimes used in flow diagrams are the sort, merge, extract, and collate outlines (see Figure 3-5). Such outlines may be used with striping, for example when a program calls a merge routine for execution from the library. For this, the predefined-process, or a striped-process, outline could be correctly shown for the call in the flow diagram, or the merge outline could be used. In a detailed flow diagram the predefined-process outline is the better choice; in a summary flow diagram the merge outline is better, since the manner of implementation is commonly expunged from summary level flow diagrams.

FLOW DIAGRAM CONVENTIONS

Striping

Striping is the use of straight lines with an outline to mark off its interior into sections. The ANSI X3.5 Standard gives a specific significance for horizontal and vertical striping within an outline. The vertical striping has already been covered in the special outline for predefined process (see Figure 4-5). Other uses of vertical striping are not specified by the Standard.

Horizontal striping is advanced by the Standard as one way of indicating a reference to another part of the flow-chart which provides a more detailed representation, as, for example, of a subroutine. A horizontal line may be drawn from the left edge to right edge in the upper portion of an outline, except for the flowline, communication link, and additional outlines. The upper area thus enclosed is used to refer to some other part of the flowchart. The lower enclosed area is used in the usual manner, as shown in Figure 4-5. Wherever a horizontal striping is used within an outline, the portion of the flow diagram referred to must, in turn, be represented on the flowchart as beginning with, and ending with, terminal outlines as described previously. Both the detailed representation and the striped outline must have location cross-references, as described below.

Cross-References

In order to make cross-referencing between parts of the flow diagram easy, two conventions are common. One is to use or to assign names to portions of the flow represented by the flow diagram. These names often are the names used to designate parts of the program. These may be the same as, or different from, the identifying names used for connectors

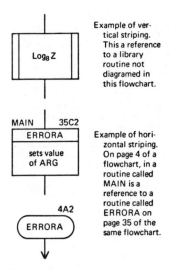

Example of vertical striping. This a reference to a library routine not diagramed in this flowchart.

Example of horizontal striping. On page 4 of a flowchart, in a routine called MAIN is a reference to a routine called ERRORA on page 35 of the same flowchart.

Figure 4-5. Conventions for striping and references

An alternative convention (not mutually exclusive of the other) is to identify a location on each physical piece of the flow diagram, as, for example, in terms of page, row, and column, as in the manner of map coordinates. An example of such a coordinate scheme is given in Figure 4-6. Another scheme that is also common is to assign a sequential number to each outline on a page following the line of flow, and then to record this number beside each outline.

The ANSI X3.5 Standard is in conflict with the ISO Standard and with previous usage in the United States on the handling of references. The ISO and general American usage has been to place the identifying name immediately above and to the left of the outline and to place the location reference above and immediately to the right of the outline. The ANSI X3.5 Standard advances exactly the opposite convention but recognizes and cites the deviation from the ISO Standard. In this book the ISO convention is used since it is also commonly used in the United States and since the ANSI X3.5 Standard explicitly recognizes the ISO position.

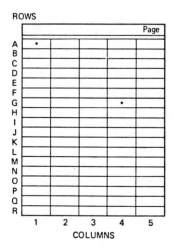

ROWS

Page

1 2 3 4 5

COLUMNS

*Example: thus, 9A1 is on page 9,
row A, column 1; and 9G4 is on
page 9, row G, column 4

Figure 4-6. One coordinate-location, cross-reference scheme for flow diagrams

Crossing Flowlines

The Standard makes specific provision for connectors and cross-references. These can be used to reduce the frequency of crossing flowlines or even to eliminate them altogether.

If it is desired to use crossing flowlines, then the convention is that the flowlines shall have no arrowheads in the vicinity of the crossover. The presence of arrowheads on the flowlines is to indicate a conjunction or coming together (fan-in) of the flow. The direction of flow at such points of joining flowlines is designated by the position and direction of the arrowheads, as in Figure 4-7.

Basic Format

Flow diagrams may be drawn with the basic and additional outlines described previously. For example, consider the program to find the average age of the employees shown in the

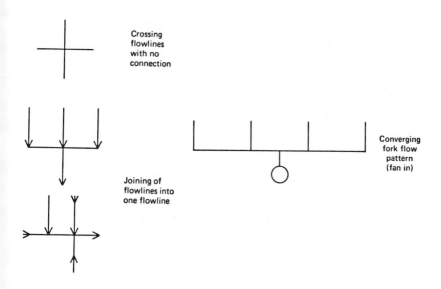

Figure 4-7. Conventions for flowlines in flow diagrams

system chart in Figure 3-16. Using the basic and the additional outlines, the flow diagram can be stated as shown in Figure 4-8.

Some comments are in order regarding this example. First, the beginning and end of the flow are marked by the termination outline. This is unlike the case of the system chart, where the start or the end of anything was an input-output outline.

Second, the sequence shown follows the common pattern of read-transform-write. Since that transformation cannot usually take place until the input data have been read, the input operation precedes the main process operation. Both precede the output operation.

Third, a very common feature of algorithms prepared for implementation on a computer is the use of iteration. This commonly appears in a flow diagram as a loop of flow. A loop is shown in Figure 4-8, but it is not the classic iterative loop. Note that in order to indicate a section of

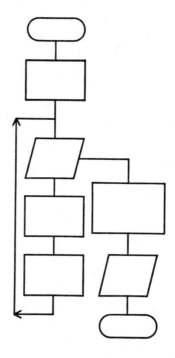

Figure 4-8. A flow diagram using basic and additional outlines

flow contrary to the normal rule, a long flowline together with open arrowheads has been used.

Just as in the case of the system chart, a flow diagram using the basic outlines alone provides some information, but not as much as it can when augmented with written identifications within the outlines. These identifications are important to indicate the portions of the data structures affected and the operations. A long sequence of process outlines which do not designate the portions of the data structures affected soon becomes ambiguous. For this reason, it is important as a practical matter to specify in detail in a flow diagram exactly what portions of the data structures are affected and in what way.

To this end, the flow diagram in Figure 4-8 can be redrawn, as shown in Figure 4-9. Here the outline choice and

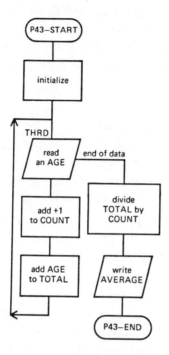

Figure 4-9. A flow diagram with identifications

sequence are identical with those in Figure 4-8, but now each of the outlines contains identifying information. The identifications consist of four things: the names of parts of the algorithm as implemented (entry points, usually, such as THRD), the names of operations (such as "add"), the names of conditions (such as "end of data"), and the names of operands (parts of the data structures affected, such as TOTAL).

Just as the person drawing a flow diagram has a choice of outlines, so also he has a choice of lettering for the identifications. One convention, illustrated in most of the flow diagrams in this book, is to put data and program names in capital letters (upper case) and operation names in lower case (small) letters. That convention probably has the highest communication value. Alternative conventions are to use upper case lettering for everything, or more rarely, lower case for everything. For computer-drawn flow diagrams, the first of these alternative conventions is the most common, since most line printers can print only in upper case.

Specialized Outlines

If one wishes to improve the communication value of a flow diagram, the usual practice is to use the specialized process outlines wherever possible. Using these outlines, Figure 4-9 can be redrawn as shown in Figure 4-10. Here the decision outline provides an explicit end-of-input-data test. The striped preparation outline refers to the short flow sequence called INITIAL.

Figure 4-10 also illustrates a use of the annotation outline. This basic outline can be helpful in the flow diagram to describe values and to provide explanation. Thus, in Figure 4-10, an annotation outline has been used to indicate, for the main process (iterative) loop, the expected number of times the loop will be executed. This information, it should be noted from Figure 4-9, is not available from the outlines or from the normal identification information supplied within the outline. Another use of the annotation outline is the warning about the need for the accurate data of the run, as shown in Figure 4-10.

Connectors and Cross-References

Connectors and cross-references are important in flow diagrams because of the common length of flow diagrams and because of convergent (fan-in) and divergent (fan-out) flows. Flow diagrams are almost always too large to represent on one sheet of paper. Usually, they include alternative flow paths. Convergence points (fan-in) and divergence points (fan-out) must be presented and the flows clearly identified.

In order to make the communication value of a flow

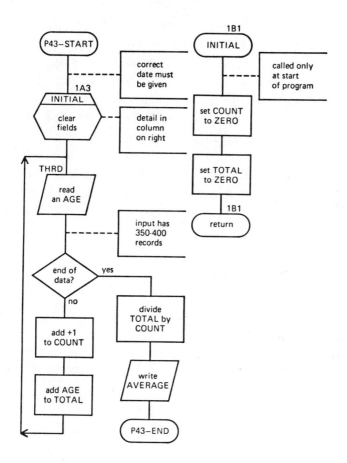

Figure 4-10. A flow diagram using specialized outlines

diagram high, one should have the flow pattern shown in as linear, or straight-line, a form as possible. The more cut up, chunky, bunched, or branched the flow pattern is, the more difficult it usually is for a person to comprehend. For this reason, the use of connector outlines and of cross-referencing normally helps to give a smoother, more linear appearance to the flow.

If the flow diagram shown in Figure 4-10 could not be represented all on one page but had to be broken into parts, it could be broken at any point. One way of doing it is shown in Figure 4-11. The connector outline does not substitute for any other outline but instead serves as an additional outline, in effect specializing the flowline symbol.

As one reads the line of flow down the page 8 part of Figure 4-11, one encounters an exit, or outconnector, labeled A3. The reader can then search Figure 4-1 for an entry or inconnector having the same identifying set of characters A3 within it. This entry or inconnector is the continuation of the flow. The identification within the entry and exit connectors must permit unequivocal and unique identification of the one appropriate entry connector to be associated with each exit connector.

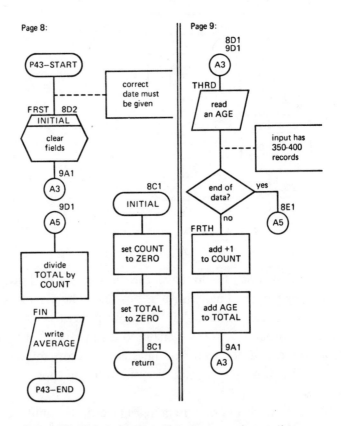

Figure 4-11. A flow diagram with connectors and cross-references

When one wants to facilitate finding the entry connector for each exit connector, the common practice is to use cross-references in the manner described previously. If any name has been assigned to a portion of the algorithm or program (that is, the program data structure), such as THRD, then this name may be entered above and immediately to the left of the connector or within the connector itself. If a co-ordinate or numbering plan has been established for identifying portions of the flow diagram, then the location indication can be entered above and immediately to the right of the connector, as shown in Figure 4-11.

The ANSI X3.5 Standard does not specify any particular coordinate convention for use specifically in flow diagrams. One common convention used is a page number followed by a row and then a column designation (see Figure 4-6 for an example). Some people omit the row designation, and some omit the column designation. Some use a sequential count to provide an index of the position within a row or within a column of a connector position. One common scheme, as also noted earlier, is to number the outlines sequentially in the line of flow on each page. Anything that begins with a digit causes no confusion since names in most programing languages may not start with a numeral.

The symbolic-name and cross-reference notations may be used with any outline, not just with connectors. They are needed for all horizontally striped outlines, as a minimum, as shown in Figure 4-11. The names also provide a way of cross-referencing any part of the flow sequence shown in the flow diagram to the actual program coding. This is illustrated in Figure 4-11 by the names FRTH and FIN, for example, which refer to portions of the program respectively, even though they are not entry points.

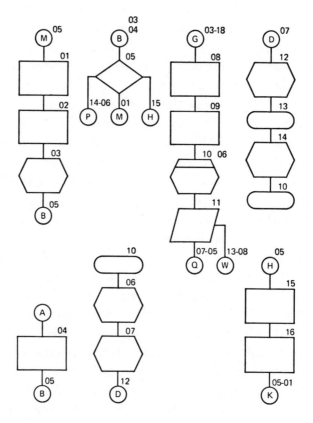

Figure 4-12. One cross-reference scheme for flow diagrams

Notation

The Standard does not specify any particular language, symbols, or notational scheme for use within the outlines to identify the data or to name operations. Ordinary English prose tends to be too verbose to be readily accommodated within small outlines. Yet using small outlines yields a more easily comprehended flow diagram. A notational scheme or set of symbols that is compact and which permits easy representation of the common situations is highly desirable.

Working toward this end some years ago, the American Standards Association (now ANSI) circulated a working paper advocating a notation or set of characters (graphics) for such a use.[1] Many users of flow diagrams prefer to use the same notation they use in the programing language. While this works well with some computationally oriented languages on computationally oriented jobs (such as FORTRAN or PL/1 in engineering or scientific work), it falls short of the need for string operations, complex operations on arrays, and for manipulations of all types of structures in logical terms. To meet these problems, several notational schemes have been advanced in the literature, of which the Iverson notation is probably the most widely known.[2]

Drawing from these three major sources, this book offers an eclectic list in Figure 4-13. This list is composed from graphics included in the ASCII, EBCDIC, and IBM BCD codes. Hence, computers can print these in computer-drawn flow diagrams. The one exception is the arrow, since none of those codes includes arrows. The arrow, however, is still widely found where communications and display equipments are used, and it has a history of use in programing work. It provides a neater alternative, especially for publication, than does a double graphic.

Of particular importance are the *is replaced by* symbols. Two alternative symbols are common and serve to indicate that the symbol on the left has its value determined by what is on the right, as illustrated in Figure 4-14. For this purpose, the equal sign has sometimes been used. Such usage is inconsistent with mathematical practice, and gives a dual role to the equal sign, the other role being to indicate equality, as, for example, from a comparison.

For comparison, a colon is common. The variable of comparison is shown on the left, and the standard or constant of comparison is shown on the right side of the colon.

[1] ANSI, "Working Paper: Graphic Symbols for Problem Definition and Analysis," *Communications of the ACM*, vol. 8, no. 6 (June 1965), pp. 363-365.

[2] Kenneth E. Iverson, *A Programming Language*, New York: John Wiley & Sons, Inc., 1962.

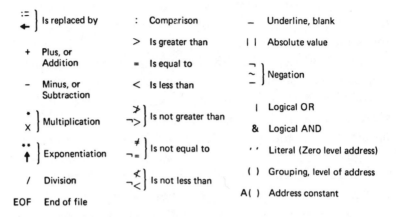

Figure 4-13. Symbols for use in flow diagrams

The exit flowlines from a decision outline must be provided with an indication of the basis for their choice, expressed in terms consistent with the notation used within the decision outline, as exampled in Figure 4-15.

Parentheses can indicate grouping, a usage borrowed from mathematics. Another usage is to indicate levels of addressing. Most literals are enclosed within prime marks (single quote marks) to indicate the zero-level addressing status. When numeric literals are to be used in arithmetic operations, they are sometimes shown without enclosing prime marks. Nonnumeric character combinations appearing without the prime marks are assumed to be first-level addresses, that is, the names of items of data, such as fields or variables.

The notation for second and higher levels of addressing is to enclose in successive pairs of parentheses, one for each additional level of addressing desired. (This and the next practice differ from that used in the early days of computers, when parentheses indicated the contents of an address.) A special variant of this is the address constant, that is, something whose value will be determined by its machine language address at the time of execution. An *A* in front of the parenthesis can serve this purpose.

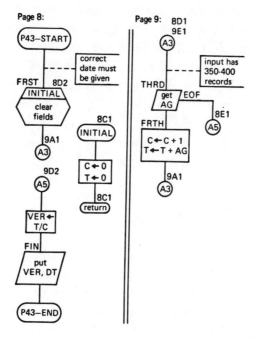

Figure 4-14. The flow diagram of Figure 4-11 redrawn, using symbols within the outlines

The use of a terse notation, such as is summarized here, permits considerably greater amounts of material to be shown within each outline, or smaller outlines, to be used in a flow diagram. In both cases there is usually an improvement in the communication value of the flow diagram.

- Do-It-Now Exercise 4.1. Using Figure 4-14 as the basis, prepare a written description in words of the operation. Be careful to be as complete and accurate as possible. Question: In what ways is the flow diagram superior in providing a means of quick, clear, comprehensive communication? What does the flow diagram leave unsaid?

• Do-It-Now Exercise 4.2. Using the specialized outlines, prepare a flow diagram to find the range in a set of numbers, assuming they are to be read one by one only once. First, use any notation you like within the outlines. Then redo the flow diagram using the notation shown in Figure 4-13. (The range is defined as the algebraic difference in value between the largest and smallest in a set of numbers. Thus the range in the set 7, 1, 4, 9, and 5 is 8.) The object of this Do-It-Now Exercise is to prepare you for the next section and the next chapter.

GUIDELINES FOR FLOW DIAGRAMS

The following guidelines have been found helpful for preparing flow diagrams. The first guideline is to chose the wording or symbols within the outlines to fit the readers of the flow diagram. This depends largely on the level of detail to be shown. The more summary (less detailed) this is, the more difficult it is to find a satisfactory wording or symbols to use within the outlines. As a general rule, whatever is chosen should be terse in order to permit the use of small outlines.

The second guideline is to be consistent in the level of detail shown in the flow diagram. If some parts of the flow diagram are in great detail and others are only sketchy, the statement of the algorithm is distorted. A consistent level of detail provides a sounder basis for making judgments about the algorithm and presents a better basis for making estimates of computer time, programing time, conversion difficulties, and debugging than does a fluctuating level. Maintaining a consistent level of detail is simple only when the level of detail matches the implementing programing language. The difficulty comes with flow diagrams at summary and intermediate levels of detail. Here it becomes a vexing problem.

The third guideline is to use identifying names consistently and to keep them brief or short. Given the type and level of symbols for use within the outline (the briefer the better), one should use names for data and operations uniformly and accurately. Figures 4-9, 4-14, and 4-15, for example, illustrate good practice.

The fourth guideline is to use cross-references liberally in the flow diagram. Cross-reference both to the program and to locations in the flow diagram improves the communication value of the flow diagram. When one wants to keep connectors small and use space efficiently, it is helpful to employ a location cross-reference inside the connector outline, and the program name cross-reference outside the connector (compare Figures 4-14 and 4-15).

The fifth guideline is to keep the flow diagram simple and clean. Clutter and lack of "white space" decreases the communication value of the flow diagram. For convenience and clarity, spacing the diagram horizontally and vertically so it can be typed (if it is prepared manually) is truly helpful in achieving a simple, clean appearance.

The sixth guideline is to keep clearly separate the operations to be performed on program-data structures from those to be performed on operand-data structures; that is, operations on the program itself, such as switch settings, indexing, initialization of program control variables, and the like, should be shown in preparation outlines, separate from operations that transform input into output data. This guideline helps make a flow diagram more easily understood, and it improves debugging work. To facilitate a clear separation, some people add a mark to the nonpreparation outlines that specify operations on the program-data structure. The mark, if used, should be one that is easily detected in a quick scan of the flow diagram. Examples are a large round spot within the outline, or a bold left edge for the outline, or a shaded upper left or other corner of the outline. None of these techniques are sanctioned by the ANSI X3.5 Standard; only the second is a violation of the Standard, and hence should not be used.

The seventh guideline is to avoid using successive connector outlines. If more than one connector should properly appear in a series (as when multiple names are assigned to one entry point, or when the program calls for consecutive, unconditional transfers of control), a good practice is to collect the connectors to the left of, or above, the line of flow. A tree arrangement (like an upside-down version of that used for multiple exits from a decision outline, as shown in the top center of Figure 4-2) can also be used.

The eighth guideline is to keep consistent the general flow pattern—from top to bottom and from left to right. When this guideline is observed, arrowheads may be omitted from the flowlines that conform to the guideline. Clarity in a flow diagram is improved by arranging the main flow to conform as much as possible to the guideline. Minor and alternative flows may then deviate from normal, and by this deviation they can be identified as not being a part of the main flow.

The ninth guideline is to draw entrances at the upper left and exits at the lower right. Entry and exit connectors are most noticeable if they are in a consistent position. Thus the usual practice is to place an entry either above or to the left of the line of flow it is to join, and to place an exit either below or to the right of the line of flow it comes from, as shown in Figure 4-15. For both cases, if the connector is set to the left or right, then the flowline goes from left to right. If the connector is set above or below, then the flowline drops down to or from the line of flow. Partial exceptions may be made to maintain symmetry, as when a decision outline has three exits—for example, exit D4 in Figure 4-15.

The tenth guideline is to draw flowlines so that they enter and exit at the visual centers of the outlines. The outlines usually possess either vertical or horizontal symmetry,

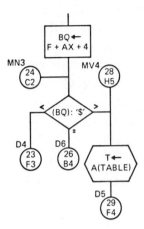

Figure 4-15. Entry and exit flowlines in a flow diagram

typically about their center points. If, therefore, entrance and exit flowlines are drawn vertically and horizontally so that they appear to point toward or emerge from the center point of the outline, the visual appearance of the flow diagram is improved.

The eleventh guideline is to use connectors and cross-references to avoid excessive crossing flowlines. Too many crossing flowlines can lose the reader in a rat's nest. How many is too many depends on the layout of the flow diagram.

The twelfth guideline is to draw only one entrance flowline per outline symbol. This is especially important in summary and intermediate levels of flow diagram. When more than one operation is specified or implied within an outline symbol, multiple entrance flowlines may mean that the sequence of events within the outline symbol is different for the respective entrances, or even that not all the operations are to be performed for one or more of the entrances. To avoid this ambiguity, one usually makes the flowlines join prior to entry into the outline, as shown in Figure 4-15. An alternative approach, lacking conformity with the ANSI

X3.5 Standard and not free of ambiguity, is to subdivide an outline.[3] When the notation within the outlines is at the same level of detail as the program itself, no practical difficulty arises. For consistency, however, this guideline should be observed at all levels of detail.

The thirteenth guideline is to draw, with four exceptions, only one exit flowline for each outline. One exception is the decision outline which by definition must have multiple exits. An example is shown in Figure 4-15. The second is the input-output outline when an end-of-file or end-of-data condition may result in a failure to read input data. The notation used with the outline should specify clearly the basis for the choice of the exit. The other exceptions are the extract and collate outlines in system charts or in summary levels of flow diagram.

The fourteenth guideline is to identify clearly all multiple exits. This is an extension of the previous guideline and is illustrated in Figure 4-15.

The fifteenth guideline is to make no violations of the Standard and to shun deviations. This guideline was explained in Chapter 3.

- Do-It-Now Exercise 4.3. Prepare a flow diagram for the following algorithm and situation: Data on body temperature recorded for a hospital patient during an exercise test are to be edited. All values less than 36.5° C. are to be replaced by 36.5° C. All values greater than 37.4° C. are to be counted. The ratio of this count to the count of the total number of temperature values is to be provided as an additional output at the end of the run. (Hint: Try modifying Figure 4-14 to cover the operations.)

---

[3] Alexandra I. Forsythe *et al, Computer Science,* New York: John Wiley & Sons, Inc., 1969.

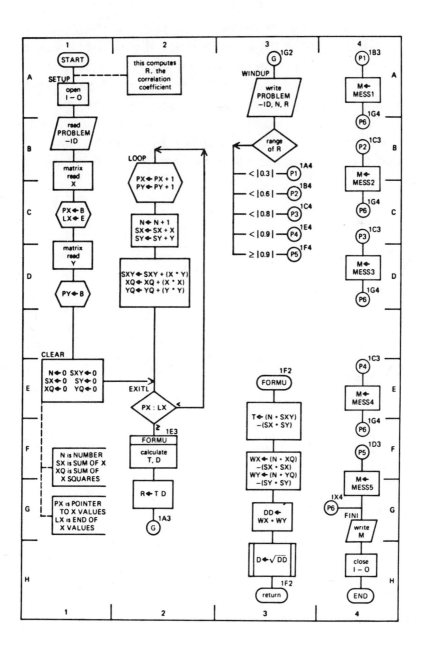

Figure 4-16. A portion of an extensive flow diagram

• Do-It-Now Exercise 4.4. Study Figures 3-18 and
4-14. Does the program P43 in Figure 4-14 produce
all the outputs needed for its role in the system
shown in Figure 3-18? If it does, write a paragraph
of explanation, using the material from the two
figures as examples, on how a flow diagram par-
ticularizes what is sketched by a system chart. If
it does not, revise the flow diagram to make it conform
with the system chart.

5

CREATING FLOWCHARTS

HEURISTICS

Creating flowcharts is an art, not a science. It is not a cut-and-dried process with firm rules. Rather, it offers the analyst and programer an opportunity to give expression to their creative urges and to apply their creative powers to problem-solving.

Creating flowcharts is closely tied to problem-solving. The problem-solving situation generally takes one of several forms: (1) Given these data, how can I have them changed to produce those data? (2) What data or transformation capability must I have to be able to get something done? (3) Why are these steps necessary—can't I replace or eliminate them with these other steps instead?

The flowchart serves in problem solving as a means of recording ideas, of summarizing possible solutions, and of stating problems in alternative forms. The flowchart serves as a vehicle for communicating ideas. This is, as pointed out in Chapter 1, the main contribution of the flowchart.

In creating flowcharts, therefore, there are two major aspects: how to use the flowchart as a communication vehicle and how to decide what to try to communicate. The first of these has been covered elsewhere in this book at some length; but the second has not.

The accumulated experience in working with flowcharts

has led to one conclusion: that creating flowcharts is a heuristic matter; that is, in creating flowcharts, people may try many approaches. Any of those approaches may fail; none can be guaranteed to yield—even under strictly defined conditions—a satisfactory solution. When an analyst or programer tries one approach and it fails to result in an accepable solution, the usual practice is just to try another approach.

Solving data-handling problems is such a varied activity that no one approach is even generally the best approach. Entire books could be written on the approaches people have tried and found successful. This book does not attempt to cover them. The focus of this book is not on how to do programing or how to do system analysis; rather, it is on flowcharts.

Fortunately, a few of the many approaches lend themselves especially well to a close tie with flowcharts. These are approaches that, by their nature, can use flowcharts as a "natural" medium of communication. Five of these approaches are presented later in this chapter. They go by various names, but are called here: known to unknown, in-do-out, eliminate assumptions, particularize, and fragment and combine. In practice, analysts and programers rarely use any approach in pure form, but mix and combine them to fit the needs they see in meeting the job at hand.

APPROACHES

Known to Unknown

The known-to-unknown approach is mostly used with system charts, not with flow diagrams. This is because the analyst or programer often finds, when he tries to describe a system, that he has available good knowledge of parts of the system, but not of the rest of it. He may not even know the full extent of the system. In such a situation, the known-to-unknown approach may serve as a convenient aid in exploring and defining the character, form, and extent of a system.

The known-to-unknown approach uses the system chart in a nearly dendritic manner. The analyst or programer starts by stating in a system chart what he believes he knows of the system. This he puts down in as much detail as possible in the system chart. Then the analyst or programer usually checks the accuracy of his system chart in order to get ready for the next step.

The system chart should consist entirely of outlines connected by flowlines. It should have no isolated groups or islands of connected outlines. If any such isolated groups exist, one step is to attempt to fill the gap. It means finding and recording the data and processes that connect the groups and close the gap. If the gap cannot be closed, then the outlying group is part of a separate system with no relation to the system under study.

The next step is to extend the system chart in both the upflow and downflow directions. On the upflow side, this means taking each input-data outline that has no incoming (entrance) flowline, and asking, "How are these data usually produced in this medium?" The answer usually permits adding at least one more layer to the sandwich. Then the same question can be asked again about the new input-data outline. On the downflow side, this means taking each output outline that has no outgoing (exit) flowline, and asking, "What happens next to these data?" The answer usually permits adding at least one more layer to the sandwich. Then the same question can be asked again about the new output-data outline.

The process can continue for a long time. A practical question is when to stop. A pragmatic answer is to stop when the processes affecting the data cannot be changed by the people who requested that a changed, or new, way of handling data be developed. A theoretical answer is to stop when the system has been completely described. The practical problem with this is that the boundaries of systems are sometimes not clearly defined.

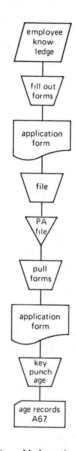

Figure 5-1. An added portion of a system

To see the use of this known-to-unknown approach, consider the system shown in Figure 3-18. Let us start with this as the known position of the system. If we look first for the upflows, two outlines are the A67 and the A34. How are the data for A67 produced? If they are punched by the personnel department from PA file documents, and if the data are entered into the file documents by the employees filling out application forms, we have traced the data to their source. This can be shown in a portion of a system chart like Figure 5-1.

The A34 data present different problems. If these data arise from any other source or are modified by any other operations that affect the age portion of the data, this can be shown. For example, if the A34 data are also subject to deletions from the processing of termination cards initiated and punched by the personnel department, then this might be shown as in Figure 5-2.

If we turn now to the output side, two outlines are the A34 and the age graph. If the graph goes to people in the personnel department for their information and filing, this

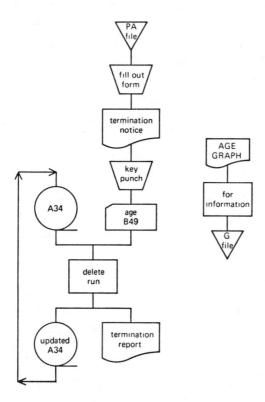

Figure 5-2. Other added portions of a system

can be shown as it is in Figure 5-2. If the subsequent uses of the A34 data do not involve the age portion of the data, no further exploration is needed. The expanded system chart can be presented as shown in Figure 5-3. Notice that Figure 5-3 also illustrates the use of lower case letters for the names of nonspecific data and operations.

Three comments about system charts are worthy of note, for they are illustrated by this example. First, the system chart does not attempt to follow the movement or flow of media, such as documents. After keypunching, are the documents refiled? What the system chart follows is the flow of the data carried by the media. Second, the system chart does not follow all uses of data exhaustively. Thus, after keypunching, the data punched are also still on the documents. The punching operation does not cause the data to disappear from the documents. The other uses of the data on the documents, even the age data, are not of concern, since they are not effective as input or output in this system. Third, the usual areas shortchanged in most system charts are in the manual and auxiliary processing of the data that are to become computer input or that originate as computer output.

In-Do-Out

The in-do-out approach is used primarily with flow diagrams, not with system charts. This is done because it does not use the sandwich structure of the system chart, but instead stresses a logical necessity for data transformations: the computer produces output data from input data. This means that the computer must take in some input data before it can produce any output data. Further, it means that the computer must do something with some input data before it can produce some output data.

This approach provides three bench marks to serve as organizing points for the operations. These points are the input, the transformation or DO, and the output. Thus, the

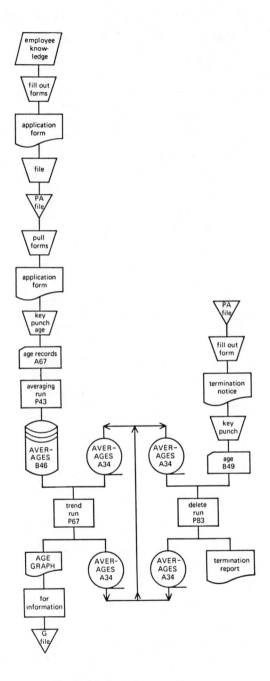

Figure 5-3. Revised system chart

analyst or the programer can start right away with a skeleton flow diagram, as is shown in the left-hand part of Figure 5-4.

Figure 5-4. Basic formats for the in-do-out approach

Then the programer or analyst can break each main outline into another set of three—setup, do, and cleanup. This means that each of the three major elements in the in-do-out skeleton can immediately be visualized in three parts: the preparation for performance or initialization, the actual performance, and the takedown or the return to normal of what may be been disturbed by the performance. This is shown in the right-hand part of Figure 5-4.

The analyst or programer then must amplify each of the outlines to show how each is to be done in the situation being studied. This requires replacing each of the outlines in the skeleton format with a sequence of outlines that reflects the circumstances of the situation being studied. It also requires giving consideration to partitioning the input into easily processed amounts and therefore producing the output in an incremented manner. This often leads to using iteration in some manner.

This in turn leads to a consideration of limits and of cycles or the position of natural breaks. These can often be built either within each other (like nested loops) or interlaced so that each adapts to the other. Examples of the former are common in the handling of the conceptual rows and columns that comprise a matrix. Examples of the latter are common in fitting output data on pages of limited size so that each page is as logically complete as possible.

Accuracy of output is of course a major concern. It requires specifying the needed operations in their proper sequence. Conservation of storage considerations leads to defining subroutines and providing calls to them. The list of possible considerations goes on indefinitely.

To see some of the simpler of these, consider the following simple problems from statistics: Create a flow diagram for a program to find the range in a set of numbers, assuming they are to be read one by one, only once; and count the number of numbers. The range is defined as the algebraic difference in value between the largest and smallest in a set of numbers.[1] Thus the range in the set 7, 1, 4, 9, and 5 is 8 (see Do-It-Now Exercise 4.2).

Using as a starting point the skeleton flow diagrams from Figure 5-4, you can make several changes. One is to look at the output. Here output is needed only when all the input

---

[1] In mathematical terms, this can be stated as

$$R = |\text{max}N_i - \text{min}N_i|$$

is complete. Hence, the output must come at the end of the read iterations. Second is the read iteration. This will require a loop in the flow and an end-of-file flow. That end-of-file flow is the place for the output, in effect as a cleanup after the reading. Drawing these yields a preliminary flow diagram such as the one in Figure 5-5.

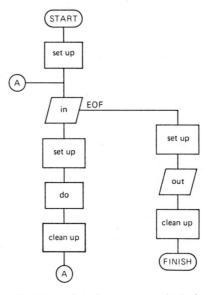

Figure 5-5. Preliminary flow diagram using the in-do-out approach

Third is the main do-part of the algorithm. For the count, a tally arrangement can be used, similar to that in Figure 4-14. From the definition of the range, one way to find the range is to compare each number as it is encountered with the smallest and largest so far encountered. If a number is more extreme than any so far encountered, then it can replace (and serve as the new) smallest or largest number. This makes the setup and the cleanup for the do very easy. Then just before output, the two numbers can be subtracted to yield the range.

This can work once the process is going, but how can the operation be started? This is an initialization problem;

because the initial values selected for the smallest and the largest must not be more extreme than the actual values that will be encountered. But what the values are is not known. One way to handle the problem is in the setup before the input. The computer could read the first number and just use it as both the largest and the smallest numbers. This requires an extra read operation—a disadvantage—as shown in Figure 5-6, and some initialization for the count.

As with all the other examples in this book, the procedure shown in Figure 5-6 is not the only way or the best way of accomplishing the data-handling. Alternatives are worth exploring, for they may offer faster execution, less use of

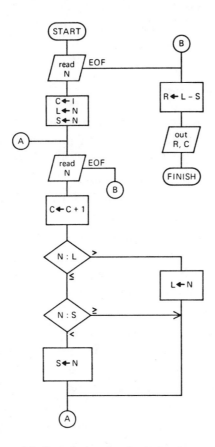

Figure 5-6. Flow diagram for the in-do-out approach

storage, simpler logic, few component operations, more convenient input and output, or less expense to prepare a run. Many of these considerations cannot be inferred easily from either a flow diagram or a system chart.

- Do-It-Now Exercise 5.1. Create again a flow diagram for the range problem, using the in-do-out approach, but also incorporate the calculation and output of the variance of the data. The variance is defined as the sum of the product of each number times itself (the sum of the squares of the numbers), minus the product of the sum times the sum of the numbers (the square of the sum) divided by the count of the number of numbers—all divided by one less than the number of numbers.

Eliminate Assumptions

The eliminate-assumptions approach is used with both system charts and flow diagrams. In practice it is one of the most widely used of all the heuristics. Often analysts and programers use it without realizing that they are using it. Some of this unintentional use results in "bugs" that later have to be corrected. A major reason for the popularity of this approach is the variety of assumptions it offers in trying to find solutions to system and algorithm problems.

This approach calls for making a series of simplifying assumptions about the job to be done. This amounts to assuming that a job is simpler than it really is. Then these simplifying assumptions are eliminated, or dropped, or replaced by less severe simplifying assumptions until the outputs and inputs conform fully to the job specifications. The best results are usually obtained by making the assumptions only about the input data and the output data. As a practical matter, assumptions about the nature of the processing or data-transformation capability available are often harder to eliminate or relax.

To see the use of this approach, consider the same problem as before, that of finding the range: Create a flow diagram for a program to find the range in a set of numbers, assuming they are to be read one by one, only once; and count the number of numbers. The range is defined as the algebraic difference in value between the largest and smallest in a set of numbers. Thus, the range in the set 7, 1, 4, 9, and 5 is 8 (see Do-It-Now Exercise 4.2).

When one uses the eliminate-assumptions approach, the first problem is deciding what simplifying assumptions to make. The starting assumptions are usually most effective if they are about the character of the input. Along this line, let us assume that the input is in ascending sorted order by the value of the number.

This assumption will help both in the initialization and in the processing of the input data. Since the first number read will be the smallest, it can be saved. Since the last number read will be the largest, it can be saved. Subtraction will then give the range. The count can be made as the new largest number is saved. This is shown in Figure 5-7.

To ease the sort assumption, you can try a range assumption. Often several assumptions must be tried to find one that can be used conveniently. One possible assumption is that the numbers are unsorted, but that the smallest is not smaller than +1 and the largest is not larger than +9. The flow diagram in Figure 5-7 can then serve as the basis for revision.

Since the two possible extreme values are known, they can be used to set the initial values of the largest and smallest number holders. Then only a single read operation is needed. But since the first number read must be used to reset the values in the two number holders, two comparisons are always needed. This is a potential waste of only a small amount

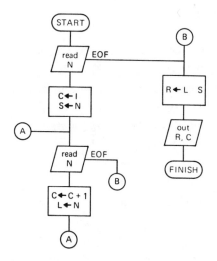

Figure 5-7. First flow diagram for the eliminate-assumptions approach

of computer time. The revised algorithm is summarized in Figure 5-8.

When you want to remove the simplifying assumption about the limits, one technique is to use the first number read as both the largest and smallest number. In order to do this with a single read, use a program switch. In the initialization, the switch can be set to "on." Next, after the read, control flows to the switch routine, which stores the number as both the largest and smallest. Then, after resetting the switch, control returns to the regular cycle. This is shown in Figure 5-9, along with the elimination of the mandatory double compare, since it is no longer needed.

A comparison of Figures 5-9 and 5-6 highlights differences only in the initialization. These differences are ineffective in altering the output—the output of both algorithms would be the same—but arise from different choices made to meet the start-up problem. The choice does not depend upon the approach used; either approach described could have led to the choice of either means shown here of meeting the start-up problem.

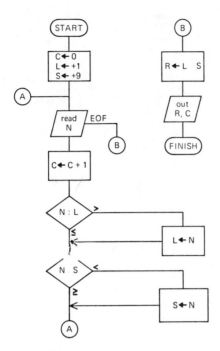

Figure 5-8. Second flow diagram for the eliminate-assumptions approach

- Do-It-Now Exercise 5.2. Create again a flow diagram for the range problem using the eliminate-assumptions approach, but also incorporate the calculation and output of the variance of the data (see Do-It-Now Exercise 5.1).

Particularize

The particularize approach is used with both system charts and flow diagrams. It is a common approach because it builds in directly the availability of a series of flowcharts from the very general to the quite detailed. Typically, programers and analysts find the communication value of such a series of flowcharts to be better than the value of a single

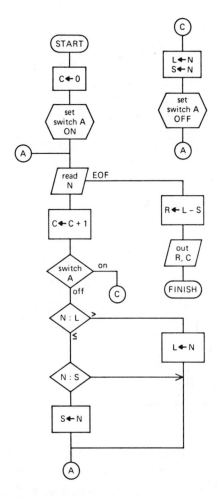

Figure 5-9. Final flow diagram for the eliminate-assumptions approach

flowchart. This is because it gives the reader greater control over the level of detail to which he wishes to expose himself.

The approach begins with the preparation of a very brief summary flowchart. The next, and each successive, step is to expand upon and amplify the outlines in the prior flowchart in order to produce a more detailed and more explicit flowchart. Some operations on data, because of their importance, typically emerge and are shown in the early summary flowcharts. Others, usually of lesser importance, appear only in the most detailed flowcharts.

An illustration of this approach may help to clarify it. For this purpose and to facilitate comparison, the example of finding the range can serve: Create a flow diagram for a program to find the range in a set of numbers, assuming they are to be read one by one, only once; and count the number of numbers. The range is defined as the algebraic difference in value between the largest and smallest in a set of numbers. Thus the range in the set 7, 1, 4, 9, and 5 is 8 (see Do-It-Now Exercise 4.2).

When one examines the problem statement, the major elements appear to be the calculation of the range and the count. This, together with the usual input and output, covers the operation, as indicated in Figure 5-10. This brief outline can serve now as the basis for a more detailed flow diagram.

Figure 5-10. Summary flow diagram for the particularize approach

Since the numbers must be read separately, it is convenient to link the count and the read together. Because the output, by contrast, comes only once, it can also be tied to the read operations for the end-of-file condition. Further, since the range cannot finally be calculated until all the input data have been processed, that calculation can also go in the end-of-file routine. This is summarized in Figure 5-11.

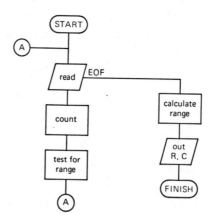

Figure 5-11. Intermediate flow diagram for the particularize approach

Since initialization is still absent from the flow diagram, clearly more detail is still to come. Yet the initialization needed cannot be completed until the manner of calculating the range is more fully specified. This, then, requires particularizing that part of the algorithm next.

The range can be calculated by using any reference point. Since the numbers will be smaller if the reference point is chosen to fall within the actual range, one manner of calculation is to measure each value against the chosen reference point and then use the largest negative and positive values. One way to get a reference point within the range is to use the first number read in the input stream. Then the difference of each successive number can be calculated from this reference number, and the greatest positive and negative differences saved, as shown in Figure 5-12.

Then, to calculate the range, it is sufficient to add together the two differences. Since *LP* must have either a

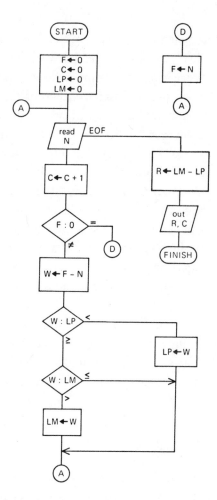

Figure 5-12. Detailed flow diagram for the particularize appoach

zero or negative value, subtracting it results in adding its absolute value. However, to establish the reference number requires either an extra comparison or another read operation or a program switch. The flow diagram shown in Figure 5-12 uses the extra comparison alternative. This, together with the other operations, requires initializing four variables.

This approach raises especially clearly a problem common to all of the approaches presented in this chapter: How much detail should be shown in a flowchart? The answer to the question depends largely on the level of detail the reader will need. Thus, if the reader will be a programer who is to prepare a program from a flow diagram, he can use any amount of detail down to that called for by the programing language to be used. In addition, he can often benefit from a more summary view. If a summary-level flow diagram, and even a detailed system chart, should be available, it could save the programer time. By contrast, a manager may only need an overview. For this, a summary-level flow diagram and an intermediate-level system chart may be adequate. In general, most users of flowcharts find they want a more detailed system chart than flow diagram, because, as illustrated in Figure 2-1, the flow diagram inherently provides more details than does a system chart.

- Do-It-Now Exercise 5.3. Create again a flow diagram for the range problem using the particularize approach, but also incorporate the calculation and output of the variance of the data (see Do-It-Now Exercise 5.1).

Fragment and Combine

The fragment-and-combine approach also is used with both system charts and flow diagrams. It finds special favor on large jobs which must be broken into parts, with the parts assigned to different teams of analysts and programers. For maintaining coordination between the teams, the usual practice is to define the interface between them in terms of what data are to serve as input for each fragment and what data are to be produced as output by the fragment.

The fragment-and-combine approach divides a job into major sections. Ideally, each section is about equally extensive and equally complex, but in practice this is extremely hard to achieve. This is partly because the actual complexity usually differs from what is anticipated and partly because the extent of the fragments is commonly redefined several times. In "modular" programing, the fragments may be made coextensive with modules, routines, control sections, or segments.

For each of the fragments, the analysts or programers then prepare a flowchart at whatever level of detail is appropriate for the need at hand. Next, combining the flowcharts for the individual fragments yields a flowchart for the entire job. To do this requires that each fragment be prepared with an eye to eventual combination. This means that consistency is needed in all fragments on names assigned to data and operations, and each fragment usually includes most of its own initialization.

As an example of the fragment-and-combine approach, consider again the range problem: Create a flow diagram for a program to find the range in a set of numbers, assuming they are to be read one by one, only once; and count the number of numbers. The range is defined as the algebraic difference in value between the largest and smallest in a set of numbers. Thus the range in the set 7, 1, 4, 9, and 5 is 8 (see Do-It-Now Exercise 4.2).

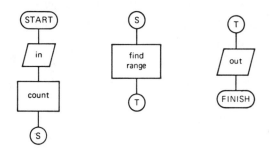

Figure 5-13. Fragments for the fragment-and-combine approach

The first step in the fragment-and-combine approach is to define the fragments. Typically, this is done from a study based on a summary-level flowchart. For this purpose, the flow diagram in Figure 5-10 can be used. Three fragments scem natural here: a read-and-count fragment, a range fragment, and an output fragment. The first fragment is to provide a count and a number. The second is to provide the range. The third is to accept the range and the count and provide them as output.

The read-and-count fragment is easily disposed of. This provides a read operation as well as a counter, which it initializes and increments, as shown in Figure 5-14. Since this fragment is also the first fragment in the flow, it traditionally must also handle the initialization of variables common (global) to several fragments. Usually these are handled in combining the fragments. The output fragment may also be easy. It need only output the range and the count, which should be provided by the other fragments if the other fragments accomplish all they are defined to do.

The range fragment is to calculate the range, given the numbers, one by one, and the count if needed. This involves an initialization problem which can be handled by a program switch. This initialization consists of setting zero as the starting value of the range R, and the first number as the largest so far encountered. The usual part of the flow must compare the number against the largest and smallest (which is equal to L-R) so far encountered. If the number is more extreme than any previously encountered, the range must be recalculated. Moreover, if the number is larger than any previously encountered, the new one must replace the old one as the largest.

The fragments can now be combined. This requires adding the switch initialization to the first fragment and annexing the output fragment to the read fragment for the end-of-file condition. The combined fragments are shown in Figure 5-15.

This approach has yielded a closely equivalent flow diagram for the illustrative problem. It differs primarily in the requirement that the middle fragment actually determine the range. Since it is not known when the last number is read until an attempt is made to read another number, the

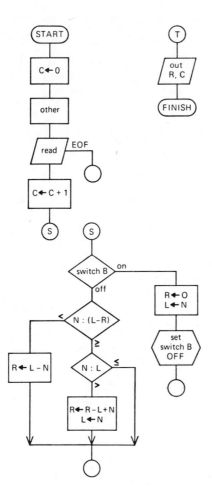

Figure 5-14. Final fragments for the fragment-and-combine approach

flow diagram must calculate the range for each number handled. This results in a longer operating time, but still only requires saving two numbers.

- Do-It-Now Exercise 5.4. Create again a flow diagram for the range problem, using the fragment-and-combine approach, but also incorporate the calculation and output of the variance of the data (see Do-It-Now Exercise 5.1).

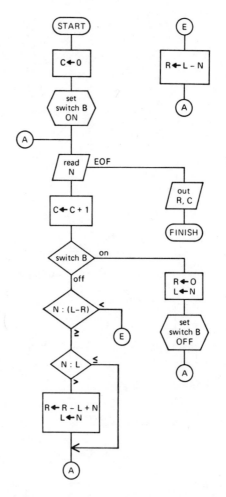

Figure 5-15. Combined fragments for the fragment-and-combine approach

PROBLEMS IN CREATING FLOWCHARTS

A major problem in creating flowcharts is the reception they sometimes get, even when well-prepared. A minority of analysts and programers hold that a flowchart is a waste of time. Since neither programers nor analysts, they claim, think in terms of flowcharts, why prepare flowcharts? Preparing and trying to read them just divert people from a more productive use of their time. This is indeed true for some people, but the documentation value of flowcharts for other people's use still justifies their preparation and maintenance.

Along this same line, others claim that alternative techniques offer superior documentation, such as, for example, decision tables,[2] logic flow tables,[3] Iverson notation,[4] networks, or directed graphs. Conceptually the ANSI X3.5 Standard could be applied to the preparation types of graphic aids other than the system chart and the flow diagram. So far, these have yet to be defined in the literature in clear form.

These objections and alternatives have indeed a measure of truth. For some situations, as noted in Chapter 1, other alternatives are superior. This can be emphasized by several weaknesses in flowcharts. First, the flowchart is weak in showing the timing of processing and of data availability or need. This lack of showing *when* is serious, for example, for communications-oriented systems and programs. Second, the flowchart does not directly tell *why;* it tells instead *what* and *how.* It leaves the *why* to human inference. Third, the flowchart does not tell *how much.* This cannot usually even be inferred. Fourth, the flowchart does not fully tell *what* or *who does* or *who is to do something* in all cases. The use of annotation is sometimes a clumsy means of trying to compensate for these weaknesses.

The second problem in creating flowcharts is how programing language is to be oriented. This is most troublesome for flow diagrams. If the flow diagram is at the same level of detail as the program and uses the same or closely

[2] Ned Chapin, "An Introduction to Decision Tables," *DPMA Quarterly,* vol. 3, no. 3 (April 1967), pp. 2-23.
[3] Sidney B. Self, "Logic Flow Table," *Journal of Data Management,* vol. 5, no. 12 (December 1967), pp. 30-36.
[4] Kenneth E. Iverson, *A Programming Language,* New York: John Wiley & Sons, Inc., 1962.

similar wording in the outlines, the communication value of the flow diagram may be only little better than a well-annotated copy of a program listing.

Further, some programing language features obscure, or make inconspicuous, certain important features of the algorithm. Thus, for example, in flow diagrams for FORTRAN and PL/1 programs, the temptation is to show DO statements in a process or preparation outline. This hides their usual loop character. For good communication, most DO statements should be expanded to assume a true loop appearance in a flow diagram. This also applies to PERFORM-VARYING statements and the like in COBOL.

The third problem in creating flowcharts is the acceptance of the ANSI X3.5 Standard. So far, even though it has existed for years, analysts and programers have largely failed to observe it. This is partly due to ignorance. The professional societies have not made its use mandatory in their publications. The computer vendors have not wholeheartedly embraced and supported it although there is some recent improvement. Moreover, the ANSI, by its refusal to grant free reprint rights and by its high charges per page for copies of the Standard, has allowed its need for income to support its work actually to result in hindering the acceptance of the Standard.

Some analysts and programers with a knowledge of the ANSI X3.5 Standard complain that the Standard is hard to use. In this regard, the two most common complaints are that the Standard wastes space and that it fails to provide enough space. People cite the "seven-page program that takes a 20-page flow diagram," and claim it is faster and easier to read the program listing. While cross-referencing in the diagram helps, it is not the entire solution. Summarization is a real solution, but this gets into level-of-conception problems as noted later.

On the charge of wasting space, many would-be users of the Standard fail to draw outlines of differing size (while preserving the ratio dimensions) to fit the wording or symbols that go within the outline. This is especially true for decision outlines and connectors as a practical matter. The two most common violations of the Standard involve the ratio dimensions of the outlines and the use of deviant outline shapes.

The fourth problem in creating flowcharts is on the level of conception. Thus it is widely agreed that a flow diagram, for example, is most useful if it is not as detailed as (is more summary than) the program it describes. To create such flow diagrams requires compressing, condensing, and eliminating details. But which ones? And how many? This is a difficult matter. A poor choice can render the resulting flow diagram nearly useless. Being more summary also increases the difficulty of providing useful cross-reference between parts of the flow diagram and between the diagram and the program. The problem is nearly as serious for system charts.

This level-of-detail problem affects the rigor and completeness of a flowchart. Prepared in full detail, everything must be present in its proper place. All tie together. When full detail is absent from the flowchart, it becomes difficult or even impossible to determine from the flowchart itself whether a particular process or operation is correctly shown or is essential. This difficulty arises particularly for decision outlines in flow diagrams and for process outlines in system charts.

The fifth problem in creating flowcharts is the burden of their preparation and maintenance. A good flowchart is no easy task to prepare. It requires a conscientious attention to detail and a fair measure of professional time and capability to create an accurate readable flowchart with good communication value. And however good it was when it was prepared, unless it is maintained to reflect currently the inevitable changes in programs and systems, it soon loses its value as it ages. The use of computers to aid in the preparation of flowcharts is a major aid, as discussed in the next chapter. However, it still requires some professional-level attention. The best results are commonly obtained when flowcharts are a part of a systematic documentation procedure that, in turn, is part of a system used by management to control and manage an organization's information-handling functions.

COMPUTER-PRODUCED FLOWCHARTS

6

BACKGROUND

Historical Development

Because of the burden of preparing flowcharts and because of the continuing need to prepare flowcharts, programers and analysts have looked for assistance. Very early, some noted that one important, possible assistance would be to have the computer help in preparing flowcharts. For computers equipped with data plotters, this seemed especially natural, but these have been much less common than line printers. As a result, most of even the early efforts to put the computer to work in preparing flowcharts used the line printers for the actual output of the flowchart.

A wide variety of programs is now available to prepare flowcharts. Most of them also produce other elements of documentation and, for that reason, are called either flowcharters or diagramers or documenters. Since these are usually available as complete software systems, their suppliers commonly market them as software packages. The suppliers include many of the computer vendors and commercial software houses, as well as individuals and major computer users.

To make sense of this variety of offerings, one should take a look at how this variety came to be. Only the highlights can be covered here, with stress placed on the development of the families of flowcharters. The published literature chronicling the development is very scant. The earliest description (1958) is of work begun in 1955 by A.E. Scott of IBM.[1] Louis Haibt's FLOWCHARTER, described in 1959, is conceptually more complex and powerful, but graphically not advanced.[2] It attempted to produce flow diagrams at different levels of detail—a difficult task.

The year 1963 saw several projects come to fruition. Don Knuth described well the advantages of using a computer to produce flow diagrams and discussed some of the implementation problems, as well as his own earlier program.[3] This apparently inspired a number of efforts, especially on the IBM-1401. Three such diagramers have since been withdrawn from public service. F. David Lewis of IBM led a project team that introduced the Symbolic Flowchart Language (SFL) and used it in the first of the series of AUTODOC programs.[4] In a parallel but unrelated development, T. J. Hamilton of IBM produced the Symbolic Block Diagraming System (SBDS). The AUTOBLOCK program, which also first saw service in 1963, was authored by D. G. Rendahl of IBM. It used as input IBM-1401 SPS; the output was in single-column form. These implementations ran on 1401 computers and were available as type III programs.

But the major development in 1963 was the 7070 AUTO-

[1] A. E. Scott, "Automatic Preparation of Flowchart Listings," *Journal of the ACM*, vol. 5, no. 1 (January 1958), pp. 57-66.

[2] Louis M. Haibt, "A Program to Draw Multilevel Flowcharts," *Proceedings of the Western Joint Computer Conference, 1959*, New York: IRE, 1959, pp. 131-137.

[3] Donald F. Knuth, "Computer-Drawn Flowcharts," *Communications of the ACM*, vol. 6, no. 9 (September 1963), pp. 555-563.

[4] F. David Lewis, "Evolution of Automatic Flowcharting Techniques," unpub. IBM internal memorandum, March 5, 1969. The author is indebted to Mr. Lewis, an early and still prominent figure in this field, for his contributions to this chapter.

CHART program (7070-AD-151), developed at IBM by a team led by Vern Mercer. This team produced a program that yielded excellent output, but placed a considerable burden on the user in terms of the input requirements.

Thus, by the close of 1963 the three major families of flow-diagram drawing programs had been founded. The basic family used the source (or object) language normally employed by the programer. A second family used the Lewis SFL input. The third family employed a cumbersome special input. (See Figure 6-1.)

Subsequent developments in 1964 elaborated each of these families. The SFL family saw the addition of 80FLOW,[5] FACTUAL, and AUTODOC-II. The source-language family saw the development of the Logic Diagramer program, which utilized the 1401 Autocoder language as the input. Three programs using modifications and augmentations of source or object languages for the IBM-1401 were produced in the 1964-1965 period, but did not prove popular. The special input family saw the development of the PRE-CHART program, which made possible a translation from 7070 Autocoder language to the input required for AUTO-CHART. This was in addition to the development of FLO-GEN, which utilized a simplified version of the input used for AUTOCHART.

In 1965 there was considerable activity, both in program development and in published papers. Prominent among the published papers were three in the *Proceedings of the Fourth Annual Meeting of UAIDE*. At that meeting Martin Goetz reported on his firm's AUTOFLOW program, which was ready for limited service.[6] It had been developed for NASA (National Aeronautics and Space Administration). G. V.

---

[5] E. Dean Houck and Louis T. Copits, "Flowcharts and Standards," *Datamation*, vol. 10, no. 6 (June 1964), p. 12.

[6] Martin A. Goetz, "Recent Developments in Automated Program Documentation." *Proceedings of the Fourth Annual Meeting of UAIDE*, San Diego, Calif.: Stromberg-Carlson Corp., 1965.

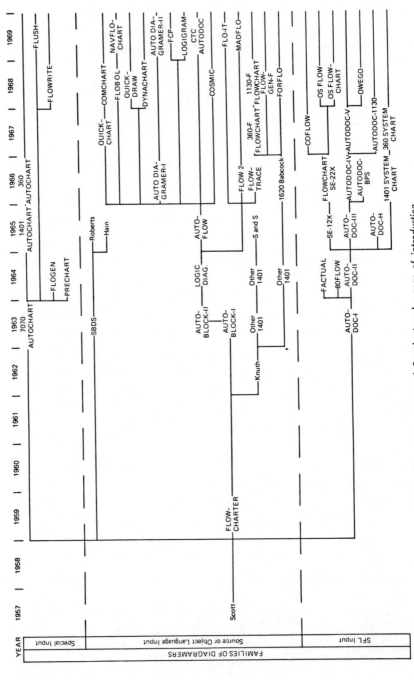

Figure 6-1. Historical summary of flowcharters by year of introduction

Roberts described his program,[7] and Saalbach and Sapovchak described theirs.[8] That same year, G. Hain and K. Hain, who also participated in the UAIDE meeting, presented a descriptive paper at the 20th ACM national meeting.[9]

Among the programs produced in the SFL family was the Documentation Aids System (SE-12X) and AUTODOC-III. In addition, a special version of AUTODOC was prepared for use on the IBM-360 at Huntsville. In the special-input family, a 1401 version of AUTOCHART was released for use.

In 1966 AUTODOC-IV and a special version of it for operations on smaller machines (BPS) were fielded, again for use at Huntsville. In the source-language family, in 1966, Philip M. Sherman described his FLOWTRACE program, developed for use at the Bell Telephone Laboratories.[10] Stelwagon fielded FLOW2 for FORTRAN, and AUTODIA-GRAMER-I took form. Robert E. Babcock introduced a program in FORTRAN for diagraming FORTRAN input for use on the IBM-1620.

In the SFL family, a new program called FLOWCHART (360-SE-22X) for the IBM-360 was fielded, which used as input a simplified and modified version of the SFL. In addition, a program to produce system charts (not flow diagrams) on both the IBM-1401 (1966) and IBM-360 (1967) was prepared and released. This was a significant first, for the main emphasis had been, and remains on, flow diagrams.

In the special-input family, a version of AUTOCHART was prepared for the IBM-360. The further advances made in

[7] G. V. Roberts, "Automated Diagram Documentation," *Proceedings of the Fourth Annual Meeting of UAIDE*, San Diego, Calif.: Stromberg-Carlson Corp., 1965.

[8] C. P. Saalbach and B. F. Sapovchak, "The Flowchart Program," *Proceedings of the Fourth Annual Meeting of UAIDE*, San Diego, Calif.: Stromberg-Carlson Corp., 1965.

[9] G. Hain and K. Hain, "Automatic Flowchart Design," *Proceedings of the ACM 20th National Conference, 1965*, New York: ACM, 1965, pp. 513-525.

[10] Philip M. Sherman, "Flowtrace, a Computer Program for Flowcharting Programs," *Communications of the ACM*, vol. 9, no. 12 (December 1966), pp. 845-854.

the special-input family in later years would draw it closer to the source-language family. This was partly because interest was renewed in the source-language family and partly because the output produced by the FLOWCHART (SE-22X) program was identical in form with the output produced by the AUTOCHART series of programs, but the input was considerably simpler.

1967 brought the release of a version of AUTODOC for the IBM-1130, as well as AUTODOC-V for the IBM-360. A translator for COBOL to SFL, named COFLOW, was produced. In the source-language family, a FORTRAN program by G. E. Gautney (360-F), FLOWCHART using FORTRAN, was released, as was QUICKCHART for Assembly Language.

In 1968 NCR's FLOWRITE was introduced in the modified special-input family. There was also the publication of Marshall Abrams' paper, "A Comparative Sampling of Systems for Producing Computer-Drawn Flowcharts." [11] In this paper Abrams compared three source-language-family programs (AUTOFLOW, FLOW2, and MADFLO, the latter being Abrams' own program) which were capable of handling the FORTRAN language. Among the other source-language-family programs fielded were the DYNA-CHART, QUICK-DRAW, FORFLO, COMCHART, FLOBOL, and FLOWGEN/F.

To provide the input for AUTODOC-V, a translator, the OWEGO, was prepared at IBM; it accepts FORTRAN as well as assembly language. In addition, two versions for use on large models of the IBM-360 were prepared for the FLOWCHART (SE-22X) program.

---

[11] Marshall D. Abrams, "A Comparative Sampling of the Systems for Producing Computer-Drawn Flowcharts," *Proceedings of the 23rd ACM National Conference P-68*, Princeton, N.J.: Brandon/Systems Press, Inc., 1968, pp. 743-750.

Recent Developments

There were no new entries in the SFL family in 1969. In the special-input family Univac presented FLUSH. This package typifies the current state of that family. The flow-diagram preparation is controlled by special notations added to the lines of source code. This resembles and approaches the convention also adopted in AUTOFLOW to force it to produce a flow diagram different from that directed by the source program.

The source-program family had a number of entries. The more prominent are noted here. Aries Corporation revised its package and offered it as AUTODIAGRAMER-II. This is one of the three packages to offer by 1970 data description as well as a flow diagram. The Navy in South Carolina revised FLOBOL and issued it as NAVFLO-CHART. This is the only major entry that produces the main line of flow from right to left across a page, instead of from top to bottom. World Computer Corporation issued FCP (Flow Chart Package), while CTSS Inc. issued the almost identical LOGIGRAM package. Both are noted for their relatively fast operation.

CTC Computer Corporation also revised and marketed Bruce Blaney's AUTODOC. This is an unfortunate confusion in names, for F. David Lewis had been using the name AUTODOC for years for the major members of the SFL family. To avoid confusion here, the CTC entry is here termed CTC-AUTODOC. This package produces a data description as well as a flow diagram featuring transfers of control, without having special notation in the source program. This, along the lines of Haibt's suggestion, is also possible with the Aries AUTODIAGRAMER-II.

Many packages introduced in previous years continue to be available. Very prominent, and accounting for a large volume of flow diagrams, is the 360 AUTOCHART in the special-input family, which is used by IBM. The OSFLOW and FLOWCHART (SE-22X) package, distributed by IBM and using a modified but SFL-based input, produce very similar flow diagrams and enjoy wide use. Also in the SFL family, the AUTODOC-IV and AUTODOC-V packages continue to be popular. More than 1,200 copies of these packages have been provided, probably a third of which are in use.

The oldest package and, partly because of that the one with the greatest number of users in the for-sale group within the source-language family, is ADR's (Applied Data Research) AUTOFLOW. More than 1,100 copies of this package have been marketed, and probably more than a third of them are in use. AUTOFLOW is available for a wider selection of computer and programing language combinations than any other package for producing flow diagrams.

The NCA's (National Computer Analysts) QUICKDRAW package has received a good market reception, in part because of its rapid speed of producing flow diagrams. The APC (Applications Programing Corporation) DYNACHART is a similar, but slower, package that devotes extra attention to presenting a logical grouping of outlines on the pages of the flow diagram. The Compress COMCHART continues active in the software market. For users of NCR computers, FLOWRITE continues to be available (special-input family).

Among packages emphasizing, or only available for, FORTRAN, the major interest has been in those for the small computers. There the IBM-1130 version by L.M. Kass, of the Gautney FLOWCHART package, has been popular. The CALCOMP (California Computer Products) FLOW-GEN/F requires a data plotter and produces flow diagrams with a single column of flow.

For producing system charts the only package available with significant capability continues to be the SYSTEMS FLOWCHART package authored by David L. Fisher. In the SFL family, it can produce system charts of any size or complexity. The package, which does not produce flow diagrams, has been available for only two models of automatic computers.

FEATURES OF PACKAGES

General Features

The packages available for producing flow diagrams have a number of features in common, but also some major differences. Many of these features can be suppressed if not wanted. As would be expected, all of the packages can produce a flow diagram at the same level of detail as the source program. All in the source-language input family can also produce a listing of the source program. Usually they tie this by cross-reference numbers to the flow diagram. In addition, all produce some type of diagnostic messages for irregularities in flow (such as exits with no corresponding entry) and some also for syntax errors in the input. These features alone certainly more than qualify the package as a means of using the computer to assist in preparing flow diagrams.

But typically the packages provide more. This *more* leads some to be advertised as "documentation" aids or systems. This *more* usually includes both a cross-reference list of all the names used to identify data and a separate cross-reference list of all the names that can be used as entry points for flow in the program. These provide a useful check on the consistency of the use of names and enhance the value of a flow diagram in debugging.

In addition to these features, most of the packages have something that sets them off from the others. Thus some, such as the CTC-AUTODOC, the AUTOFLOW, and the Aries AUTODIAGRAMER-II, feature a data description in addition to a data cross-reference. Some, such as the QUICK-

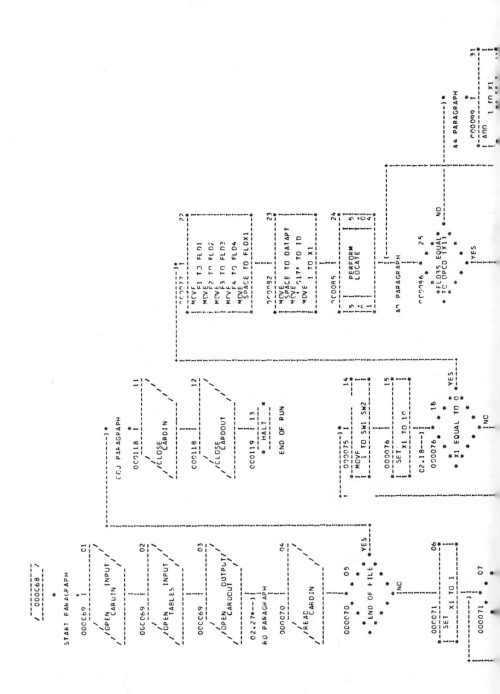

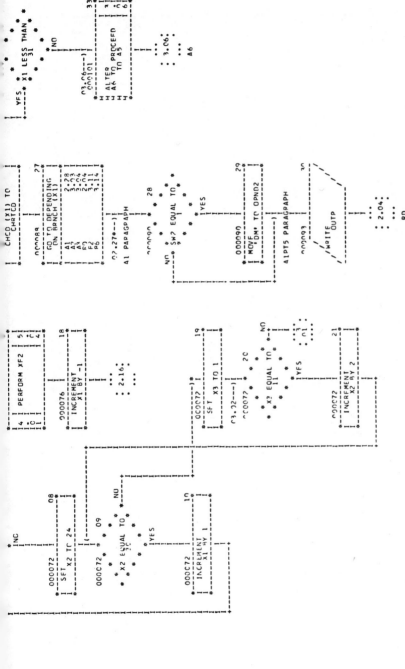

Figure 6-2. Example of a part of a flow diagram produced by ADR-AUTOFLOW

NAT'L. COMPUTER ANALYSTS, INC. QUICKDRAW   CROSS REFERENCE TO DATA NAME   PAGE 1

```
         DATA NAME    SEQ.              SEQUENCE NUMBER OF REFERENCES
        BRNCH        0062  *  0088 0119
        CARDIN       0018  *  0069 0070 0116
        CARDOUT      0032  *  0069 0116
        CH           0053  *  0116 0108 0139 0145
        CHCD         0061  *  0087
     -  FLDX1        0139  *  0081
        FLD1         0055  *  0077
        FLD2         0038  *  0078 0111
     -  FLD3         0040  *  0079 0075 0107 0112 0138
        FLD35        0041  *  0086
     -  FLD4         0047  *  0080 0110
     -  F1           0024  *  0077
        F2           0025  *  0078
     -  F2A          0026  *  0132 0137 0141 0144
     -  F3           0027  *  0079
        F4           0028  *  0080
     -  F4A          0029  *  0154 0150 0151 0152
     -  ID           0054  *  0083 0143
        INPT         0023  *  NOT REFERENCED
        OPCD         0059  *  0086 0119 0119
        OPND2        0048  *  0090 0096 0108
     -  OP1          0043  *  0104
        OP3          0042  *  0103
     -  OUTP         0037  *  0093
        SPEC         0052  *  0095 0113 0148
     -  SWITCHES     0063  *  NOT REFERENCED
        SW1          0064  *  0075 0133 0134 0141
        SW2          0065  *  0075 0090 0134 0151 0152
        TABLE        0057  *  NOT REFERENCED
     -  TBLENT       0058  *  NOT REFERENCED
     -  TOP          0050  *  0105
        X1           0066  *  0071 0021 0076 0076 0084 0086 0287 0088 0097 0098 0119 0119 0132 0134 0135
                              0137 0141 0144 0146 0150 0150 0151 0152
```

Figure 6-3. Example of the data names cross-reference produced by NCA
QUICK-DRAW

DRAW, feature speed of execution. Others such as the AUTOFLOW feature easy adaptability or availability for different computers and languages, while still others, such as the AUTODOC-V and OSFLOW, carry a price tag (they had been available free).

Specific Features

In their details the packages differ on a number of features that are not obvious upon a casual inspection. Consider, for instance, the format of the flow diagram. Most packages produce a 22-inch vertical depth of diagram on 14⅞-inch-wide paper. This is equivalent to a size twice that of the page size most commonly used in line printers. Across this double page the packages normally produced either three, four, or five columns of flow (outlines). The IBM-distributed and AUTODOC-V packages produce five columns. Most of the others produce either three or four columns, also in a fixed position on the page. The AUTO-FLOW produces from one to four centered columns of flow. The CALCOMP FLOWGEN/F produces one column on 8½×11-inch-sized paper. The NAVFLOCHART packages produce a fixed number of rows.

The factors that limit the capacity of the packages differ. In some, as has been true of the AUTOFLOW, the limitation is in the number of pages of output that can be produced; in others it is the number of data and program names; in a few others it is the number of statements in the source program.

In general the capacity factor appears not to be a major limiting consideration. But approaching the limit of any of the packages results in slowing down the performance through a decreasing efficiency of external storage use. As a practical matter, most of the packages appear designed for efficient operations with less than 400 names, with source programs of less than 1,500 lines, and with output amounting to less than

PAGE 1

TABLE OF CONTENTS AND REFERENCES
AUTOFLOW CHART SET - SAMPLE

CARD ID   PAGE/BOX   NAME

REFERENCES (SOURCE SEQUENCE NO. AND PAGE/BOX)

COBOL MODULE

CHART TITLE - REMARKS

CHART TITLE - PROCEDURE DIVISION

CARD ID	PAGE/BOX	NAME	REF 1	REF 2	REF 3	REF 4
(000179)	2.01	START				
(000182)	2.08		(000182) 2.10			
(000183)	2.11		(000182) 2.08			
(000187)	2.14	CLEAR-STAT	(000182) 2.09			
(000188)	2.15		(000182) 2.09			
(000190)	3.01	READ-TRANS	(000183) 2.11 (000266) 6.01	(000188) 2.15 (000299) 7.14	(000193) 3.04 (000322) 8.13	(000262) 5.17
(000203)	3.11		(000200) 3.09			
(000206)	3.12	EOF-TRANS	(000192) 3.03	(000194) 3.05	(000205) 3.14	(000261) 6.01
(000204)	3.14	READ-TRANS-EXIT	(000183) 2.11 (000299) 7.14	(000193) 3.04 (000322) 8.13	(000262) 5.17	
(000329)	3.15	EOJ	(000207) 3.12	(000234) 4.12		
(000333)	3.19		(000333) 3.21			
(000334)	3.22	EOJA	(000333) 3.20	(000333) 3.19		
(000334)	3.22		(000333) 3.20			
(000351)	3.29					
(000210)	4.01	READ-MASTER	(000184) 2.12	(000263) 5.18	(000272) 6.05	
(000222)	4.09		(000219) 4.07			
(000224)	4.11		(000224) 4.18			
(000233)	4.12	EOF-MASTER	(000212) 4.03	(000232) 4.20		
(000225)	4.14	COUNT-OLD	(000224) 4.17			
(000225)	4.14		(000224) 4.11			
(000229)	4.19	COUNT-OLD-A	(000226) 4.14	(000228) 4.16	(000228) 5.01	

(000231) 4.20 READ-MASTER-EXIT	(000184) 2.12	(000236) 4.13	(000263) 5.18	(000272) 6.05
(000228) 5.01	(000227) 4.15	(000224) 4.17	(000273) 6.05	(000300) 7.14
(000237) 5.02 COMPARE	(000186) 2.13 (000323) 8.13	(000264) 5.18		
(000241) 5.05 EQ	(000238) 5.02			
(000245) 5.08	(000243) 5.06			
(000260) 5.15	(000253) 5.12			
(000265) 6.01 DELETE-MASTER	(000244) 5.07			
(000268) 6.03 HIGH	(000239) 5.03			
(000274) 6.06 LOW	(000240) 5.04			
(000276) 6.08	(000276) 6.10			
(000324) 6.11 CARRWALK-ERROR	(000275) 6.06			
(000277) 6.13 LOW-SCAN	(000276) 6.09			
(000277) 6.13	(000276) 6.08			
(000327) 6.16	(000325) 6.11			
(000328) 6.17	(000326) 6.12			
(000280) 6.18	(000276) 6.09	(000279) 6.14		
(000281) 7.01 LOW-A	(000278) 6.13	(000280) 6.15	(000280) 6.18	
(000292) 7.06	(000285) 7.03			
(000294) 7.09	(000294) 7.11			
(000295) 7.12 LOW-SCAN-A	(000294) 7.10			
(000295) 7.12	(000294) 7.09			
(000296) 7.13	(000294) 7.10			
(000298) 7.14 LOW-B	(000296) 7.12	(000296) 7.13		
(000301) 8.01 WRITE-IN-ERROR	(000242) 5.05			
(000303) 8.02 ERROR-PRINT	(000328) 6.17			
(000321) 8.12	(000319) 8.08	(000320) 8.10		
(000352) 8.14 END	(000335) 3.22	(000351) 3.29		

Figure 6-4. Example of the program names cross-reference produced by ADR AUTOFLOW

30 double pages of output diagrams. The program with the highest capacity limit is the IBM FLOWCHART (22X) program, which can accommodate without modification up to about 9,000 names.

With regard to the conformance of the flow diagrams to the ANSI X3.5 Standard, a wide diversity of deviation can be observed. Only one of the packages available in 1969 conformed closely to the Standard, the IBM FLOW-CHART (22X), and that only when a special option is invoked. The Lewis AUTODOC-V package can be used in a manner that approaches conforming to the Standard, but such conformance depends on the human user, not on the package itself.

The most frequent violation of the Standard is in the ratio sizes of the outlines. The Standard specifies a specific ratio of width to height. The only programs not seriously violating this portion of the Standard are those programs which provide only one size of outline and which draw in the five-column format.

The Standard specifically provides for changing the size of an outline provided that the width-to-height ratios be maintained. This is not a conceptually insuperable problem and could be handled in a way that conforms to the Standard. Yet all of the packages that permit a variation in the size of an outline, limit the maximum width, while they may place no limit (such as AUTOFLOW), or an arbitrary fixed limit (such as DYNACHART), on the height. The result can be a severe distortion in the configuration of the outlines.

To avoid partially this difficulty, some packages (such as AUTOFLOW) generate annotation to carry the wording that could otherwise force an outline beyond a recognizable shape. Others truncate the wording either partially or after some limit on the number of characters printed. Others deliberately make a distortion in an outline in order to accommodate the wording. For example, the CTC-AUTODOC provides four sizes of decision outlines. If even the largest is too small, the package "drops the bottom" of a portion of the outline and extends the wording in a column of print.

Connectors are a common violation. Typically, programers want the connector outlines small, but the data names to go within them, not small. To get around this problem, the packages use a variety of devices, none of which conforms to the Standard. Entry connectors are particularly deviant, but exit connectors and terminal connectors also show considerable deviation. Some even omit connector outlines entirely, either in all places, or in selected positions in the flow diagram.

Another major violation of the ANSI Standard is in the handling of annotation. The Standard indicates it should be handled by means of an annotation outline. In practice, none of the packages does this in the way indicated by the Standard. Some write the notes in running text form as a break in the flow, some enclose them in a rectangle, some place them with top and bottom lines in or adjacent to the flow, some optionally suppress them, and others optionally provide for increasing or adding to the notes.

Since annotation of the source listing is considered one of the major ways of documenting a program, the lack of conformance to the Standard on annotation might be rationalized on the grounds that such annotation is appropriate in the source listing rather than in the diagram. This is because the diagram by its graphic format provides additional aid to the reader and hence need not rely on the annotation.

But if annotation has value in a source list, it should also have value in a diagram, the counter argument runs. This is because the purpose of the diagram is to communicate as much as possible about the program; hence, the more annotation, the better. But even this argument cannot be a defense of the packages' nonconformance to the ANSI Standard on annotation.

Still another violation is the use of non-Standard outlines.

Some people regard these as deviations or augmentations of the Standard since they typically are used for specific situations for which the Standard specifies a general outline.

A related matter is the handling of cross-references. Some packages provide a means of using the programer-assigned numbers (page and line) for the statements that comprise the program or of using sequence numbers generated by the package itself. The latter gives a more consistent means of reference, but requires that the package also produce a source listing which shows the numbers it assigns to the statements. Such numbers can then be used in the cross-reference listings and in the flow diagram to go from any name or outline back to the source listing.

Obviously this can work only if the flow diagram does indeed incorporate the cross-reference numbers. Most packages put the numbers above and to the left of the outline, whereas some put them above and to the right and some do not use them at all. Further, it can only work if the cross-reference lists use the numbers, too, as most of them do.

To provide cross-reference from the source listing of the program to the flow diagram, two means can be used. One is to use the cross-reference lists to find the name desired. Since these lists are usually in a sorted order by the name, the procedure is usually easy. Then the reader can use the sequence number shown in the table and scan the flow diagram for the numbers that are close in sequence. This can be a frustrating chore.

To ease the chore, some of the cross-reference lists also incorporate a page-and-outline indication in the cross-reference. While most of the packages do this for the names of entry points in the program, few (among them, AUTO-FLOW) do it consistently for both data names and program entry-point names. For this purpose, the packages usually number consecutively in the line of flow all of the outlines on a page. The packages then print this page and outline or just the outline number usually above and to the right of the outline or sometimes above and to the left of the outline. A few do not print them at all on the flow diagram.

These page and outline numbers can then be used to indicate locations in the flow diagram. Thus they can be used as the wording within connectors or used with an exit connector to indicate the location of the corresponding entry connector. Some packages do not use them at all with outconnectors when the entry connector is on the same page as the exit connector. These page and outline numbers can also be used with the entry connectors to show the locations of the corresponding exit connectors. Here the packages differ considerably. Some, like AUTOFLOW, cite only the first, with an asterisk to indicate that there are more than one. Some, like DYNACHART, cite all.

Another use of the location indicators is for subroutine calls, for DO's, and for PERFORM's. Here the usual practice is to show the location of the start and often also the end of the part of the program to be executed. Even packages that normally omit locations from the flow diagrams, like the CTC-AUTODOC, include them here since such information is not easily obtained from the cross-reference lists. Some packages, such as AUTOFLOW, include a table summarizing them.

Still another aid to cross-reference is the inclusion of the program-entry names directly in the flow diagram. The practices used by the packages do not conform to the ISO or ANSI Standard on this although the QUICK-DRAW comes reasonably close. Others put program-entry names in breaks in the flowlines, or to the left of the flowline. Some also put them either above or below a location connector. Their presence on the flow diagram helps speed the cross-reference to the source program.

Related to the position of connectors is the extent to which the package rearranges the source program as it produces the flow diagram. Most packages take the source program in the order in which it was written and produce a diagram in substantially that order. A few, such as the AUTOFLOW, may sometimes attempt to rearrange, or to move to more logical positions, selected portions of the source program. Others, such as the DYNACHART, attempt to fit routines and named portions of a program on a flow-diagram page and attempt to start new pages with entry points whenever possible.

One other feature of the flow diagrams produced by the packages is not at all obvious—the total amount of output produced by the package for each 100 imperative statements in the program. For some, such as the numbers of the SFL family (like AUTODOC-V), this is largely controllable by the programer. For most, it can be varied by the optional features selected or suppressed. To some extent, however, it is a function of the package itself. Some packages such as the LOGIGRAM, the AUTODOC-V, and the DYNA-CHART produce relatively thin output, while the CTC-AUTODOC and the AUTOFLOW produce a relatively copious output. For all the available features, the difference amounts to more than 50 percent in terms of total pages of output. This greater output volume also affects the speed of execution of the packages.

- Do-It-Now Exercise 6.1. Take a short program that you know or understand. From the source listing, prepare a detailed flow diagram in the best form you can, on large size (14⅞×22-inch) paper. Then run the same program against a flowchart-producing package. Compare the computer-produced flow diagram with the one you produced. Make a list of all the points of similarity and difference. Which flow diagram is better, and why?

SELECTING FLOWCHART PACKAGES

When one selects a package for producing flowcharts with the help of the computer, several considerations appear especially important. These and a few of the less significant matters are presented here briefly, in a rough order of practical importance:

1. *Will the package run on the configuration?* This depends on the peripherals available, the amount of storage, the usual operating system, the model of the computer, and the computer's command repertoire. The difference between what is possible and what is convenient also should be noted. When a choice exists, the best configuration to use is the one that has magnetic disks and is most accessible to the programers and analysts.

2. *Will the package handle all the major languages used in the facility?* One that handles only COBOL may serve COBOL programers well, but may be useless to PL/1 or assembly-language programers. The dominant computer language at the facility should get the most attention. Packages for some languages, such as SNOBOL, LISP, APT, and APL, are not available commercially, and it seems unlikely that the common packages will be revised to provide for such less common languages.

3. *Is support readily available?* Bugs are almost certain to appear in the use of any major program package. That a bug occurs is a poor reflection on the supplier of the package. What counts is the speed and adequacy of the correction of the bug. This requires support from the supplier. The support may also include or exclude updates or modifications to the package. The policy on this should be clearly understood. The support, in practice, depends in part on geography.

4. *Is the package easy to run?* This consideration is not just for the common cases, but also for the instances when someone wants something not commonly asked for, yet pos-

sible with the package. This reflects both the training provided by the supplier and also the adequacy of the documentation materials covering the package.

5. *Can the package fit well into my facility's documentation practices?* This is a question of integrating the use of the package into the management practices used at the facility. If the use of the package can contribute to and further the management's achievement of the facility's objectives, then the package will be more valuable and better used than if its use is somebody's "crackpot idea."

6. *Does the package have the desired features?* Nothing is going to please everyone, but the wider the possible selection of features, the more likely the package is to be accepted.

7. *Do the package's flow diagrams conform to the ANSI X3.5 Standard?* There is no reason why the flow diagrams should not conform and also no reason to accept a package with serious violations of the Standard.

8. *Is the cross-referencing good?* This is one of the most important features of the flow diagrams that the packages produce because of the extensive level of detail they include. The cross-references help to enable the user to find his way around among the details.

9. *Is the flow diagram easy to read?* This is a subjective matter, but one that influences the communication value of the flow diagram.

10. *Is the price reasonable?* Packages are available at a wide range of prices. What is reasonable depends on the features, the support, and the competitive values, as well as the circumstances of the facility.

INTERPRETATION AND USE OF FLOWCHARTS

Once a package has been selected for producing flowcharts with the help of the computer, the question then becomes one of how to get the best use from it. The following suggestions have been culled from some of the experience of users thus far. Fortunately, with only a little difference in stress, these suggestions serve as guides to the use and interpretation of flowcharts generally, however prepared.

1. *Use the flowchart for communication.* After what has been said earlier in this book, this may seem almost trite; but it is not, for the *how* is sometimes elusive. First, a flowchart can serve as a memory aid. It relieves an analyst or programer of the need to try to keep many details in mind all at one time; that is, the flowchart communicates for one person, from one time to another. Second, a flowchart can serve as a work-defining aid. It communicates between different people about what is to be done or has to be done. To these ends, a flowchart should be regarded, treated, and used as a major means of conveying ideas.

2. *Mark up the flowchart.* Use it as a piece of note paper to highlight ideas, to note things to do, to explain why something was done. Treat it as you would working scratch paper.

3. *Don't save flowcharts.* Properly used scratch paper has an element of timeliness about it. After it has become stale, its proper home is the wastebasket. Many flowcharts should go there too, but some will carry notes that should be transcribed or written up to become part of the documentation for a system or program before the flowchart goes to the wastebasket. Rather than work to keep a flowchart current and in step with a program or system, the best policy, especially for flow diagrams, is periodically to create updated ones from the source programs in the system. Let the programs and procedures be the base; let the flowcharts just be descriptions. One of the advantages of the flowchart packages is the ease of producing an up-to-date flowchart at any time, quickly and fairly easily.

4. *Cut, tape, and paste with flowcharts.* By their manner of creation, flowcharts are fitted on pieces of paper of limited sizes. These have no relationship at all to the program or system. Hence, to get a really useful picture of a

program or system from a flowchart, one must eliminate the page size limit. As a practical matter, this means some cut, tape, and paste work to put the flows where they logically belong. It also means drawing some long flowlines to show relationships more clearly.

5. *Replace nearby connectors with arrow flowlines.* The connectors are easier to draw, especially in the computer-produced flow diagram. Yet the arrow flowlines typically are more easily read. As long as a rat's-nest effect can be avoided, the communication value of a flow diagram is enhanced by the use of arrows. They are often especially effective on a cut, tape, and paste flowchart.

6. *Highlight the main flows.* Redraw or mark up the flowchart to make the main control paths and flows appear as long, unbroken chains. Figure 3-17 gives this idea. For good judgments about a program or system, the flowchart must show by its structure what is of main importance and what is of lesser importance.

7. *Concentrate on logic first.* The major problems are usually with the organization and structure of data handling. The petty problems are usually with the details of the means of accomplishing the data handling. In practice this usually means first paying attention to the things that determine the main flow paths through the flowchart. To highlight these, a good practice is to mark them (as in color) or to redraw the flowchart to concentrate in one part of the flowchart the operations that determine the main flow paths. This organization of the flowchart enhances its communication value.

8. *Group operations into related modules.* Grouping keeps like operations together and hence makes them easier to comprehend and work with. The criteria used to group the operations may be quite complex and involve elements not directly shown on the flowchart, such as time factors. Grouping also helps in assigning people to work on the program or system and helps to relate work done on another part.

To indicate the grouping in the flowchart, one must mark it or redraw it.

9. *Keep the size of the flowchart outlines small.* This suggestion can be implemented by drawing the flowchart in the usual manner, but on 14⅞×22-inch or 17×22-inch paper, and then passing it through a size-reducing copier. The 8½×11-inch result is still readable and, in the case of computer-prepared flowcharts, may even be more readable. The small-size flowcharts are physically much easier to work with.

10. *Use cross-references and annotation.* Both cross-references and annotation can supply data that otherwise are unavailable and can provide additional depth of information, amplifying and clarifying what is told by the main body of the flowchart. Their role and value has been pointed out earlier in the interpretation of flowcharts.

7

DRAWING FLOWCHARTS

SITUATIONS

In the previous chapter we presented information on drawing flow diagrams with the aid of a computer. When the packages are available this is a significant time-saver for analysts and programers when the program exists in a source language. The typical result, as noted earlier, is a detailed flow diagram produced with wording borrowed from the source program.

Chapter 2 includes directions for the use of templates for drawing flowcharts and contrasts them with the freehand drawing of flowcharts. As a practical matter, templates are widely used for drawing neat, readable flowcharts. Unlike the computer-drawn flowcharts, those drawn with the aid of templates have continuous lines and can be either system charts or flow diagrams.

Not every situation can be handled well by using any of these three techniques for handling the mechanics of drawing flow diagrams. For example, sometimes outlines must be made neatly, but also larger or smaller than the openings provided in the template. Or, for example, a draftsman may have to prepare a flowchart in a special layout of large size for subsequent photo reduction. The usual practice for meeting these situations is to draw the outlines directly.

This is not easy, if the ANSI X3.5 Standard is to be observed fully.[1] It requires a technical description for each outline in terms strict enough for an artist or draftsman to use. Yet the description should be understandable enough for the average programer and analyst to enable them to draw the outlines too. The remainder of this chapter attempts to provide such a description.

TERMS

Ratios. The dimensional ratio of the outlines defined in the ANSI X3.5 Standard is determined by the following procedure. Construct a rectangle circumscribing the outline. The rectangle must be formed from vertical and horizontal lines, and each line should just touch the inscribed outline. The ratio is then determined by measuring the horizontal and the vertical sides of the circumscribing rectangle; the ratio is usually defined as the width taken as unity, to the height taken as a fraction of the width. The metric and tenths-of-an-inch scales on the S and P templates are helpful for these measurements.

Sizes. The Standard permits outlines of any size as long as the dimensional ratio and general configuration are maintained.

Lines. The Standard indicates the use of any uniform width or weight of line for all of the outlines, regardless of their sizes. Lines need not be continuous, but may be created by the close spacing of discrete symbols, as in computer-produced flowcharts.

Configuration. The Standard indicates that the outlines in shape should conform closely enough to the specified configurations to permit rapid and unambiguous identification by a user of the flowchart. The curvature of lines and the

[1] ANSI, *Standard Flowchart Symbols and Their Use in Information Processing (X3.5)*, New York: American National Standards Institute, 1971.

magnitude of angles may vary from those shown in the Standard provided that the shapes are still clearly recognizable. To that end, since the angles and curvatures shown in the Standard are sometimes difficult to use in drawing, the configurations and description given in this chapter sometimes simplify or round off to make the drawing easier, within the restrictions imposed by the Standard.

Orientation. The figures in this chapter illustrate the general orientation specified by the Standard for each outline. Outlines may not be turned. Each outline, except the connector and the decision outlines, has at least one straight line which must be either vertical or horizontal, as illustrated in the figures. The connector outline has no specified orientation; the decision outline is horizontally oriented along its greatest dimension. Flowlines should normally be vertical or horizontal and may make bends. Flowlines deviating from the horizontal or vertical are neither recommended nor proscribed by the Standard.

OUTLINES

Input-output. The input-output outline is a parallelogram with its base edge about 15 degrees further to the left than is top edge. This cuts off about one-sixth from the top and bottom edges. The outline has a width-to-height ratio of 1 to ⅔ as illustrated in Figure 7-1.

Process. The process outline is a simple rectangle with a width-to-height ratio of 1 to ⅔ as shown in Figure 7-1.

Flowline. A flowline may be of any length. The spread between the barbs of the arrowhead and the length of the arrowhead are each about ten times the width of the flowline as shown in Figure 7-1. The angle of the barbs is 26½ degrees from the flowline; that is, the entire arrowhead just fits within a square.

Annotation. The annotation outline is a rectangle with

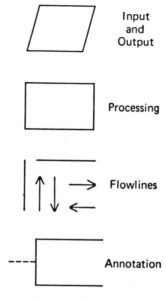

Figure 7-1. Basic outlines

the right side missing, but with a width-to-height ratio of 1 to ⅔ as shown in Figure 7-1.

Connector. The connector outline is a circle, as shown in Figure 7-2. In practice, it should be kept as small as possible. A size just large enough to accommodate four characters of wording is about typical.

Terminal. The terminal outline has a width-to-height ratio of 1 to ⅜ and looks like a flattened ellipse or a slim rectangle with half-circle ends, as shown in Figure 7-2.

Parallel mode. The parallel-mode outline consists of two parallel lines of any equal length, spaced about 10 line-widths apart as shown in Figure 7-2. Entering the upper or left one, there may be one or more flowlines positioned anywhere along the line. Leaving the lower or right one, there may be one or more flowlines positioned anywhere along the line. Usually the outline is oriented horizontally with vertical flowlines. The number of entrance and exit flowlines may not both be equal to 1.

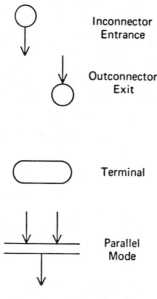

Inconnector
Entrance

Outconnector
Exit

Terminal

Parallel
Mode

Figure 7-2. Additional outlines

Document. The top three edges of the document outline are a portion of a rectangle, but the bottom edge is a curved line to represent a break in the paper. The width-to-maximum-height ratio is 1 to ⅔, as shown in Figure 7-3. The width-to-height ratio of the rectangular portion is 1 to ½. The radius of the left curve is one-half of the width, and the center is one-fourth of the width in from the left edge. The radius of the right curve is one and one-fourth of the width, and the center is straight below the right edge of the outline.

Magnetic tape. The magnetic tape outline is a circle with a horizontal line tangent to, and to the right from, the bottom. The tail line, as shown in Figure 7-3, extends over to the point of intersection with an imaginary vertical line tangent to the rightmost edge of the circle.

Punched card. The punched-card outline has a width-to-height ratio of 1 to ½ and appears generally like an upper left corner-cut punched card, as shown in Figure 7-3. The corner cut has an angle of about 30 degrees and cuts off about one-sixth of both the width and height.

Punched tape. The punched-tape outline has a maximum width-to-height ratio of 1 to ½, as shown in Figure 7-3. The centers of the curves are at the one-fourth points of the width; the radii of the curves are three-fourths of the width. The end lines are vertical. The top left corner and the bottom right corner are not more than 10 percent of the height in from imaginary horizontal lines tangent to the maximum points on the arcs.

Display. As shown in Figure 7-4, the width-to-height ratio of the display outline is 1 to ⅔. The radius of the curved lines is one-half of the width. The neck curves join the horizontal lines one-third of the width in from the left ends.

Manual input. The trapezoidal manual-input outline, as shown in Figure 7-4, is a stylization of a cross-section of a keyboard with the sloping surface (at an angle of about 10 degrees) of the keyboard having its lowest point to the left.

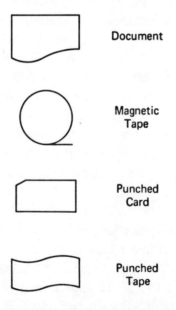

Document

Magnetic
Tape

Punched
Card

Punched
Tape

Figure 7-3. Specialized outlines for media

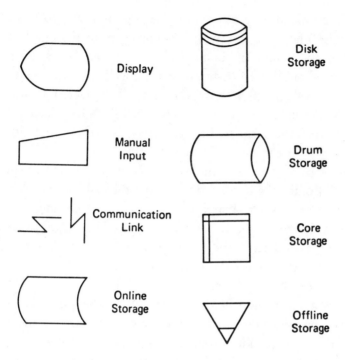

Figure 7-4. Specialized outlines for equipment

This cuts about one-third off the height of the left vertical edge. The width-to-maximum-height ratio is 1 to ½.

Communication link. As shown in Figure 7-4, the zigzag of the communication-link outline has an angle of about 45 degrees. The distance between the parallel line segments is from 10 to 15 times the width of the flowline. The lines may be of any length, and one or more zigzags may be located anywhere along the lines.

Online storage. The online-storage outline is a stylization of a portion of a cylinder with a convex end at the left and a concave end at the right. It has a width-to-height ratio of 1 to ⅔, as shown in Figure 7-4. The ends are arcs with a radius equal to one-half of the width.

Disk storage. The disk-storage outline shown in Figure 7-4 is a reorientation of the online-storage outline, but with three lines added, one convex on the top, and two to mark off bands. These are arcs with a radius equal to the one-half width of the equivalent online-storage outline. The spacing of the bands is about one-tenth to one-twelfth of the width of the equivalent online-storage outline. The overall width-to-height ratio is 1 to 5/3. The ANSI X3.5 Standard has indicated the ratio to be ⅔ to 5/4, but that seems inconsistent with the outline's configuration.

Drum storage. The drum-storage outline as shown in Figure 7-4 omits the two band arcs from the disk-storage outline, and reorients it to match the online-storage outline. The overall width-to-height ratio is 5/3 to 1. The ANSI X3.5 Standard has indicated the ratio to be 5/4 to ⅔, but that seems inconsistent with the outline's configuration.

Core storage. As shown in Figure 7-4, the core-storage outline is a square and hence has a width-to-height ratio of 1 to 1. The two lines, each parallel to a side, are in about one-eighth of the width from the edge of the outline.

Offline storage. The offline-storage outline is an equilateral triangle standing on a point. Since it is equilateral, it has a width-to-height ratio of 1 to 0.866 (about 15 to 13), as shown in Figure 7-4. The small line drawn about six-tenths or eight-thirteenths of the distance from the top to the bottom tip of the triangle is a required part of the outline.

Decision. As shown in Figure 7-5, the decision outline is a diamond outline with a width-to-height ratio of 1 to ⅔.

Preparation. The horizontal lines of the preparation outline, as shown in Figure 7-5, are two-thirds of the total width. This makes the angle of the four sloping sides fall at 26½ degrees from the vertical, since one-sixth of the total width is missing from each end of the horizontal lines. This gives a width-to-height ratio of 1 to ⅔.

Predefined process. The predefined-process outline is a rectangle with vertical bars in about one-eighth to one-sixth of the width from the left and right ends. Figure 7-5 shows

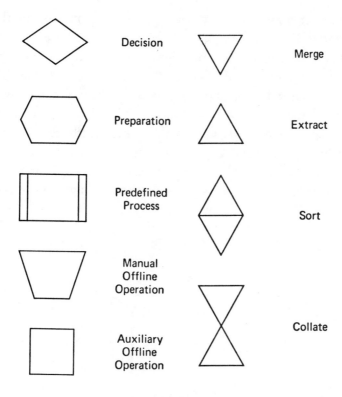

Figure 7-5. Specialized outlines for processes

the lines at about the one-eighth position. The overall shape is the same as the basic process outline with a width-to-height ratio of 1 to ⅔.

Manual process. The manual-process outline is a keystone-shaped (trapezoidal) outline, as shown in Figure 7-5, with a maximum width-to-height ratio of 1 to ⅔. The slope of the sides is about 15 degrees. This makes the length of the bottom horizontal line equal to one-half of the width.

Auxiliary process. The auxiliary-process outline, as shown in Figure 7-5, is a square.

Merge, extract, sort, and collate. The merge, extract, sort, and collate outlines are shown in Figure 7-5. All are constructed from equilateral triangles. Those that use two abutting triangles must use triangles of the same size for the two parts.

Card deck. As shown in Figure 7-6, the card-deck outline is an extension of the punched-card outline. The left corner outline is extended to give a height about one-fifth greater than the punched-card outline. The topmost horizontal line is extended about one-eighth of the width beyond the right end of the punched-card outline embedded in this outline, and the new right vertical is made slightly shorter, as shown in Figure 7-6. This gives an overall dimensional ratio of about 5/4 to ⅔ for width to height.

Deck of cards

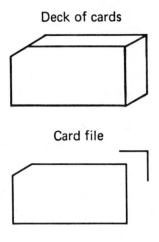

Card file

Figure 7-6. Specialized outlines for punched-card media

Card file. The card-file outline is like the card-deck outline, but with parts of the top, left slant (corner cut), right slant, and right vertical erased. The position of the top and right lines is determined in the same manner as for the card-deck outline. The remaining upper-right right angle has lines equal in length to about one-half of the height of the punched-card outline embedded in this outline.

INDEX

173